Farmers, Subalterns, and Activists

There is growing global interest in chemical-free sustainable agriculture—and not just for ecological reasons. Advocates claim that sustainable farming methods challenge the power of agri-business over food systems and empower rural communities. Yet, in countries such as India, this challenge to power is paradoxical. Organisations that promote sustainable agriculture in India claim to challenge dominant powers in the agrarian sector, but in order to achieve their goals they must retain close relations with powerful individuals and institutions. Most rely on transnational capital to fund their operations. To achieve results on the ground, they often ally themselves with state governments and rural elites. For political representation, they rely on middle class activists who often have more to gain by aligning themselves with trending development discourses, rather than truly representing the views of the poor and marginalised. Through the social relations they forge, sustainable agriculture organisations are woven into the fabric of hegemony.

Farmers, Subalterns, and Activists explores how this paradoxical position of sustainable agriculture organisations plays out on the ground in rural India. It provides detailed case studies from three regions, which illustrate the various ways in which organisations oppose the dominant food order, but also how they become ensnared within it, and the implications of this for their work on the ground.

For scholars, the book presents fresh perspectives on agrarian change and the ways in which power infuses sustainable rural development efforts. For development practitioners, it provides an easy-to-read demonstration of why social power matters in efforts to promote sustainable agriculture. It argues that practitioners must consider the landscape of social power when developing rural development projects and listen to the needs and perspectives of the poorest and most marginal within communities if their interventions are to be effective, equitable, and sustainable.

Trent Brown is Australian Research Council DECRA Fellow in the School of Geography at the University of Melbourne. His research explores various aspects of social transformation in contemporary India, including rural development, urbanisation, youth transitions, skill development, and sustainability issues.

Farmers, Subalterns, and Activists

Social Politics of Sustainable Agriculture in India

Trent Brown

CAMBRIDGE
UNIVERSITY PRESS

CAMBRIDGE
UNIVERSITY PRESS

University Printing House, Cambridge CB2 8BS, United Kingdom

One Liberty Plaza, 20th Floor, New York, NY 10006, USA

477 Williamstown Road, Port Melbourne, vic 3207, Australia

314 to 321, 3rd Floor, Plot No.3, Splendor Forum, Jasola District Centre, New Delhi 110025, India

79 Anson Road, #06–04/06, Singapore 079906

Cambridge University Press is part of the University of Cambridge.

It furthers the University's mission by disseminating knowledge in the pursuit of education, learning and research at the highest international levels of excellence.

www.cambridge.org
Information on this title: www.cambridge.org/9781108425100

First published 2018

Printed in India by Thomson Press India Ltd.

A catalogue record for this publication is available from the British Library

ISBN 978-1-108-42510-0 Hardback

For my mum and dad

Contents

List of Images

Acknowledgements

This book is informed largely by my doctoral research, undertaken at the University of Wollongong from 2009 to 2013. Fieldwork in India was made possible through financial support received from the Centre for Asia Pacific Social Transformation Studies (CAPSTRANS) and the Faculty of Arts at the University of Wollongong.

I would like to thank my supervisors Richard Howson, Susan Engel, and Tim Scrase for their valuable guidance, suggestions and criticisms. I benefitted tremendously from having input from three scholars with diverse areas of expertise—in sociological theory, the political economy of development, and South Asian studies, respectively.

I am particularly indebted to Kanchi Kohli, from Kalpavriksh, New Delhi, whose input was especially helpful during the formative stages of the research. Kanchi's support was extremely valuable in developing a research focus, identifying issues of local significance, and connecting me with relevant people in India.

During my fieldwork in India, I had the pleasure of meeting a number of scholars, who received me with great hospitality and were a source of great intellectual stimulation. In particular, I must acknowledge Manjit Singh (Professor of Sociology at Panjab University, Chandigarh) and Shubhprem Brar (Professor of Geography at Rajindra College, Bathinda), who added a great deal of depth to my perspective on Punjab. Manjit was of great help in getting me established in Punjab and helping me to develop a broad perspective on the history of social movements in India. Shubhprem, to whom I feel especially indebted, was an insightful guide as I conducted my research in Punjab, teaching me a great deal about the geography, culture, and political economy of the region. I look back with great affection on the many, thought-provoking discussions we had regarding my research and a range of local issues.

I extend my thanks to Angrej Singh Gill and Deepak Singh Rotella, who assisted me as translators in Punjab and Uttarakhand, respectively.

I have had the good fortune of receiving academic input from a number of sources, but I am particularly grateful to Tanya Jakimow and Michael Gillan, who both took the time to provide detailed and thought-provoking suggestions on papers I presented at conferences. Both helped me greatly in developing the analysis that is presented in this book.

I would like to thank Brian Martin, for organising the High Output Writing Programme at the University of Wollongong, which helped me to develop a habit of regular writing. I am also grateful to all of those who participated in the weekly meetings of the programme (too numerous to name individually) and provided useful feedback on sections of this book.

I would also like to extend thanks to Ruchira Ganguly-Scrase, who contributed to an engaged research culture amongst South Asian studies scholars at the University of Wollongong through organising seminars, reading groups, and film screenings. I thank Bokhtiar Ahmed, Emily Pleskun, Andrew Deuchar, Susheela Pandian, Vicki Crinis, and George Matheson, who were all regular participants within this research community.

I owe a special debt of gratitude to Raj Patel, who examined my PhD thesis. In his comments, he suggested that I make hegemony a more central focus of my analysis—a suggestion which I took up in revising my thesis and even more so in putting together this book.

After completing my PhD thesis, I undertook follow-up research in India, which I have incorporated in this book. This included a period in early 2015, when I revisited Tamil Nadu for a series of additional interviews. During this time, I received research assistance from A. K. Arockiasamy, to whom I am deeply grateful. I also subsequently received assistance from Susheela Pandian, who helped with Tamil translations.

In late 2015, I presented an overview of the findings of my doctoral research at the Centre for South Asian Studies at Kyoto University (KINDAS). Scholars at Kyoto University provided a series of constructive criticisms, which I have attempted to take on board while reworking my PhD thesis as a book. I thank them for their input.

I must thank Kuntala Lahiri-Dutt, Craig Jeffrey, and Amanda Gilbertson for guidance through the process of converting my thesis into a book format. I am also grateful to the three anonymous reviewers of this book manuscript, who provided valuable suggestions to sharpen the focus of the book and to Anwesha Rana and Anushruti Ganguly at Cambridge University Press, for their support throughout the publication process.

Finally, I would like to thank all of those who gave up their time to participate in my research. In particular, I would like to thank Umendra Dutt, M. Revathi, R. T. Thiruvenkataswamy, Sudesha Devi, Vijay Jardhari, and Biju Negi, who were extremely generous and patient in providing me with access to their organisations, giving interviews, providing local connections, and allowing me to stay in their homes as I conducted my fieldwork. Although the tone of some sections of this book may appear critical towards aspects of their work, I nonetheless found their dedication, passion and zeal to be genuinely inspiring and wish them every success in their valiant attempts to forge more sustainable food systems.

1

Introduction

The twentieth century brought unprecedented changes in the agrarian structure of many so-called developing societies. Major interventions into farming technologies and practices, the social relations of agricultural production, and the arrangement of global food distribution systems saw the emergence of what McMichael (2013) refers to as the 'corporate food regime'. This refers to the global constellation of food and farming systems, which has been organised in such a way as to maximise opportunities for capital accumulation for large, multinational corporations and transnational finance, often through the exploitation and liquidation of labouring agrarian classes.

The Green Revolution, a capital-intensive rural development model introduced throughout the developing world from the 1940s to the 1970s, provided a key building block for this corporate food regime. In its endeavour to rapidly increase food production through use of high yielding variety (HYV) seeds, chemical inputs, modern irrigation systems, and mechanisation, the Green Revolution generated immense profits for transnational agri-business, whilst having mixed results for farmers and labourers on the ground. This greatly expanded the influence of transnational capital in third world agriculture, providing a basis for the increased corporate control and financialisation of agriculture from the 1970s onwards, ushering in the 'corporate food regime' in its current form (Patel, 2013).

The current social condition of rural societies throughout the developing world testifies to the major shortcomings of the Green Revolution and the corporate food regime. The Green Revolution imposed one-size-fits-all technologies that failed to adjust to the diverse social and ecological conditions at the grassroots (Altieri, 1984). As a result of side effects such as environmental pollution, increased vulnerability to pests and disease, declining soil fertility, and growing indebtedness, the yield increases observed in the early stages of the Green Revolution have, in many cases, not been sustained, leading to widespread economic stagnation in the rural societies of the Global South (Conway, 1985). This issue has been compounded by the growing influence of agri-business and finance over the food system, which have undermined farmers' autonomy and introduced price instability, rendering rural livelihoods highly precarious (McMichael, 2013). In the future, there are further problems projected for this corporate-controlled farming system. As Altieri and Toledo (2011) note, the practice of monoculture within Green Revolution-style agriculture leaves it vulnerable to collapse in the face of climate change, and its dependency on high-energy inputs makes it vulnerable in an energy-constrained future.

In India, the language of 'crisis' is frequently invoked in academic discussions of contemporary rural society (Lerche, 2011; Patnaik, 2003; Reddy and Mishra, 2009a; Walker, 2008). This crisis can be thought of as having two key moments. The first relates to the after-effects of the Green Revolution. Today, agricultural production in regions that followed a Green Revolution development path have stagnated, while input costs have increased, reducing farmers' income, often forcing them to indebtedness (Singh, 2009). Furthermore, the resource-intensive approach has had adverse ecological effects, including a decline in the water table (Rodell, Velicogna and Famiglietti, 2009) and an increase in environmental toxicity (Tiwana *et al*, 2009), which has, in turn, had adverse effects on human health (Kochhar *et al*, 2007; Thakur *et al*, 2010). The second moment in the development of agrarian crisis has been the effects of liberalisation on India's rural economy. Liberalisation has constrained access to rural credit, removed subsidies for inputs, and lowered food prices, severely diminishing the viability of farming as an economic venture (Patnaik, 2003). As the growth in urban employment opportunities in India has severely lagged behind the loss of livelihoods in the countryside (Reddy and Mishra, 2009b), rural populations are in an increasingly desperate position. In this context, many small-holding farmers have been forced into debt to remain afloat, while the corresponding stress of indebtedness has been implicated in the massive increase in farmer suicides in India since liberalisation took effect (Sainath, 2009).

These developments render the current food regime 'unsustainable', in the sense that it does not provide rural populations with the means of subsisting in the long term. It is in this context that farming systems that reduce production costs, decrease ecological and health burdens, and offer means of value adding may be seen as 'sustainable' and desirable alternatives.

The Promise of Sustainable Agriculture

There are clearly many different models for more sustainable food systems. Some suggest that, given declining agricultural efficiency and population pressures across many developing countries, the only sustainable path forward is a shift in populations from rural to urban areas and a corresponding transition to a more industrial agricultural model. Others favour technological 'solutions', such as the introduction of genetically modified crops, to provide a boost to farm outputs. The agricultural models that I refer to as 'sustainable agriculture' in this book, however, are best described as 'low external input sustainable agriculture'. In some respects, this is an umbrella term, encompassing a variety of farming models, such as organic farming, natural farming, and biodynamic farming, all of which have emerged in response to the problems of industrial and Green Revolution agriculture, described previously. What these models have in common is that they focus on ecologically integrated techniques and the development of local food markets to reduce farmers' dependency on market inputs that are external to the local farm system—such as chemical fertilisers, patented seeds, and heavy machinery—theoretically reducing farm expenses whilst lowering the ecological burden of agriculture.[1]

There is now a large network of organisations spread throughout the world that suggest that such models of agriculture offer a way out of the current agrarian crisis experienced in India and other developing countries. Instead of bringing in high-energy inputs from outside, sustainable agricultural models draw and build on local ecological cycles to meet crop requirements. This provides obvious solutions to the environmental side effects of industrial agriculture, having less chemical pollution, less entropy, and lower resource intensity. Importantly, it also provides solutions to the socio–economic crises of industrial agriculture. By lowering the reliance on external inputs, which, increasingly, only large-holding farmers can afford, it is argued that sustainable farming systems reduce the expenditure and indebtedness of communities, particularly small holders (Uphoff, 2002). By enabling farmers to develop the self-sufficiency of their farming systems, they foster greater autonomy from the dictates of global agri-business and the corporate food regime (Holt-Giménez and Altieri, 2013). Furthermore, in being relatively more labour-intensive than industrial, chemical farming, they provide livelihood opportunities for the labouring rural poor (Pimentel, 1993). At its best, sustainable agriculture is a dynamic phenomenon, which seeks to address a diverse range of problems—not only ecological, but social, economic, and cultural—to build farming and food systems that can endure in the long term.

Yet, advocates of sustainable agriculture often go beyond stressing these technical merits: they claim that the farming systems they promote offer a pathway to a reconfigured, more socially just food regime, within which social power and hegemony are radically transformed. It is this claim which this book explores. The goals of farmer autonomy, localised and non-exploitative distribution systems,

[1] An important distinction here is between low external input sustainable agricultural models, and those that rely on organic 'input substitution' (Rosset and Altieri, 1997). In the latter, synthetic chemical inputs are simply replaced with other forms of 'organic' external inputs, which may allow the food to be certified as 'chemical-free' for consumers, but still rely on ecologically and socio–economically unsustainable practices.

healthier consumption patterns, and more integrated, harmonious communities are major themes in both the literature and practice of the sustainable agriculture movement worldwide. The realisation of these goals seems predicated on the assumption that those marginalised by the corporate food regime—smallholding farmers, the landless, women, tribal populations, and so on—will rise to positions of leadership within the movement and work to develop new farming systems that better reflect their needs and aspirations.

The question this book poses, however, is whether this is how the movement for sustainable agriculture operates in practice, or whether, instead, the movement comes under the influence of other groups and institutions with interests in maintaining the status quo. Throughout the course of this book, I critically interrogate the claims of scholars and activists that sustainable agriculture, as it is implemented on the ground, represents a meaningful challenge to hegemony and social relations of power. I do this by examining how sustainable agriculture organisations sit in relation to hegemonic forces. With a specific focus on India, I show that while sustainable agriculture organisations are often motivated by a counter-hegemonic ideology, in practice, they must work through a web of hegemonic power relations that criss-cross the countryside, which can thwart their potential to challenge the status quo. I thus bring together two seemingly disparate fields: sustainable agriculture's claims to social justice and the analysis of hegemonic relations in rural India.

Certainly, there is clear evidence from various parts of the world that sustainable agriculture at least has the *potential* to challenge power relations. As an ostensibly pro-poor project that challenges agri-business dominance and the logic of neoliberalism, sustainable agriculture presents as a counter-hegemonic project. Its counter-hegemonic potential is most apparent in Latin America. In parts of Latin America, widespread use of biodiverse, non-commercial agriculture has been documented. Communities engage in these forms of agriculture even when they come with a significant economic opportunity cost, as they provide food security, protection from market instability, personal satisfaction, and a sense of community and cultural identity (Isakson, 2009). In doing so, they allow communities to build their self-reliance and resist the risks associated with full incorporation into the global market economy (Isakson, 2009). It is not surprising, therefore, that sustainable farming practices have become integral to Latin America's radical peasant movements, including the now globally active social movement La Via Campesina (Martínez-Torres and Rosset, 2010). Indeed, Altieri and Toledo (2011) describe a 'revolution' in agro-ecology in Latin America, within which small-holding peasants have taken a lead role. With their practice of resource-conserving methods and knowledge of local ecologies, smallholders have provided a voice for the ways in which farming can be done sustainably, to reduce costs, gain autonomy from agri-business, and meet the nutritional needs of local communities. Frequently, knowledge regarding the practice of sustainable farming and its empowering potentials has been distributed through autonomous, decentralised farmer-to-farmer networks (Holt-Giménez, 2006). In some of the socialist Latin American countries, these perspectives have been taken up by the state as part of strategies to encourage food self-sufficiency. For several years now, ecological technologies, local production (including urban farming), and land reform to promote biodiverse smallholder production have been key components of the agricultural policies of Cuba (Rosset and Bourque, 2005) and Venezuela (Cunich, 2012).

For these reasons, sustainable farming practices are regarded as central to the concept of 'food sovereignty', the central organising principle of La Via Campesina. The concept of food sovereignty has resonated broadly throughout the world and has been adopted by a wide variety of farming organisations (Fairbairn, 2012). As a counter point to the more mainstream concept of 'food security', food sovereignty emphasises that, in addition to having the right to access affordable, safe, and healthy food, communities have the right to define the conditions under which they produce food. Sustainable agriculture is said to do this, by reducing expenses on external inputs and encouraging greater on-farm crop diversity.

Yet, in India—as well as much of developing Asia—the scenario is quite different and considerably more ambiguous to that in Latin America. Sustainable agriculture has generally not been propagated in and through pro-poor peasant movements. Rather, it has been articulated through Non-Government Organisations (NGOs) and other institutions that are thoroughly woven into existing hegemonic relations. They are often led by urban middle-class activists or rural elites who have historically played the role of articulating hegemony and establishing capitalist relations of production in the countryside (Deshpande, 2003; Byres, 1981). They rely on transnational donors for financial support, many of whom are amongst the chief agents who have propagated neoliberalism in the developing world (Petras, 1999). Furthermore, they rely on the state for ongoing support, and the Indian state, at least since the 1960s, has shown a clear commitment to promoting capitalist, intensive, technology-heavy agriculture (Varshney, 1995). In this sense, sustainable agriculture in India is somewhat paradoxical in its relation to hegemony: ideologically, it is opposed to the hegemonic logic of neoliberalism, the Green Revolution, and the corporate food regime; yet, in practice, it is closely tied to the very institutions that have propagated the established hegemony and have strong interests in maintaining and expanding the existing food regime.[2] By

[2] It may be said that a similar paradox could be detected in a vast array of development NGOs operating in the Global South—not just those promoting sustainable agriculture. This may

taking up case studies to examine in close detail how this tension plays out on the ground, this book provides a timely reflection on the strengths and limitations of sustainable agriculture in challenging and reproducing hegemonic constellations in contemporary rural India.

India's Sustainable Farming Movement: A Brief Overview

Current civil society initiatives for sustainable agriculture in India have complex historical roots. Some suggest that the work done by Gandhian rural collectives before Independence could be regarded as a form of organic farming (Scialabba, 2000). Until the 1980s, however, most efforts were highly localised and uncoordinated. Claude Alvares (2009: 65–67), who has been at the forefront of India's organic movement for more than thirty years, has suggested that a turning point came in the mid-1980s and 1990s, when civil society groups with an interest in sustainable agriculture began forming national-level networks which provided a basis for knowledge sharing and collaborations. These nascent networks benefitted from access to experimental agricultural plots at the Auroville commune in Tamil Nadu, where international experts in sustainable farming had been active for several years. In the early 2000s, Alvares formalised these networks under the banner of the Organic Farming Association of India (OFAI). OFAI facilitates contact between natural and organic farmers throughout the country with regular meetings and engages in much more concerted initiatives for providing outreach and support. It is a relatively professionalised network with regional secretariats co-ordinating activities throughout India. In addition to OFAI, there are also a number of issue-specific national networks that formed in the early 2000s. The Millet Network of India, for example, which formed in 2008, is a coalition of NGOs that promote the use of millets as a hardy, drought-tolerant, and climate-change resilient crop that provides numerous health benefits compared to wheat or rice.

These initiatives provide a national platform for networking, knowledge exchange, and advocacy—yet, the vast majority of individual civil society organisations for sustainable agriculture have a regional or state focus. These organisations are diverse in their structure and mode of operation. There are many large internationally-funded NGOs working on sustainable agriculture that have been very successful in deploying their resources to reach large numbers of farmers. The Centre for Sustainable Agriculture in Hyderabad, for example, has partnered

well be the case. Nonetheless, I would suggest that this paradox is particularly pertinent to the study of the sustainable agriculture movement, as its ideology often makes strikingly counter-hegemonic claims, and the hegemonic institutions that it must work both with and against are numerous.

with a number of smaller NGOs to distribute knowledge on non-pesticide management strategies. Their efforts have been highly successful, with 350,000 farmers using these techniques in Andhra Pradesh as of 2007 (Ramanjaneyulu *et al*, 2009), a figure that has reportedly increased substantially since then.

There are also several noteworthy initiatives that have attempted to incorporate sustainable agriculture within a broader platform for social change. The Honeybee Network, for example, which has been active since the late 1980s, aims to harness the grassroots innovations of rural communities, augment them with contemporary scientific knowledge, and provide opportunities to upscale these innovations to develop commercial rural livelihood options. The Honeybee Network encourages not only rural development but also an epistemic shift—to recognise the rural poor not simply as 'beneficiaries' of innovations originating elsewhere, but as important innovators in their own right, as exemplified through sustainable farming techniques (Gupta, 2013).

In addition to these high-profile organisations are a large number of more informally structured groups that do not receive enormous funding from external agencies. There are a number of local initiatives that simply aim to raise awareness and exchange knowledge on sustainable techniques (Pastakia, 1998). There are also a large number of small seed-saving collectives throughout India, which have taken up the task of conserving genetic diversity in 'seed banks' (Vijaylakshmi and Balasubramanian, 2004). In some cases, these networks for seed saving, awareness generation, and knowledge exchange are organised as social movements and incorporate a campaigning element into their work. The case studies presented in this book fall into this broad category.

Many Indian sustainable agriculture advocates express themselves in an ideological idiom that claims to be uniquely Indian. India's agrarian history and identity is used as a way of valorising the transition from chemical to sustainable farming. The connection between the farmer and nature is depicted as a quintessentially Indian relationship. As the respected natural farmer Bhaskar Save (2008: 6) puts it, 'Farming runs in our blood'. The emphasis within India's sustainable agriculture movements on the use of *panchagavya* and other bio-inputs derived from cow urine (Natrajan, 2008) highlights India's unique relationship with the cow. It also highlights the effectiveness of traditional techniques that are indigenous to India, having been developed over thousands of years. There are also discernible links between the discourses that these activists use and Indian philosophical traditions, both ancient and modern. There is certainly a widespread recognition that sustainable agriculture accords with Gandhi's vision of development and several leading figures within the movement have acknowledged Gandhi as a chief source of inspiration (Save, 2008: 33–34). Gandhi's philosophy of national development took as its starting point agriculture and village life

(Gandhi, 1963). He advocated the development of each village as a self-sufficient economic unit, without any dependency upon external resources. Shortly after Independence, Gandhi's disciples, J. C. Kumarappa and Mira Behn, were among the first to oppose the use of chemicals in agriculture in India, highlighting the damage it would cause to soil and its impact on India's balance of payments (Guha, 2007: 223–224).

In the post-Green Revolution period, Gandhi's ideas of village self-reliance (*swaraj*) took on a new significance and have been adopted by many figures within the sustainable agriculture movement. Subhash Palekar (n.d.), for example, whose books on the philosophy and practice of sustainable farming have become highly influential in India, argues that the external inputs introduced with the Green Revolution created a system of dependency and exploitation. Consequently, for Palekar, the ideal agricultural system, which he calls 'spiritual farming', should entail no external inputs of any kind—whether chemical or organic. The farming methods that he advances thus rely exclusively on materials available at the village level. Shripad Dabholkar (2001) takes these 'localist' ideals even further, outlining methods of harvesting the maximum amount of energy and knowledge available within local agricultural and social systems to create a more just and prosperous society. On the basis of his own experiences in rural Maharashtra, Dabholkar has developed a series of techniques for facilitating local, grassroots movements, with a focus on presenting knowledge of sustainable techniques in a language that is accessible to those without formal education. Though expressed in an idiom that purports to be 'Indian', the ideals of farmer independence that underlie these articulations have strong resonances with global 'food sovereignty' discourses.

There have also been attempts to emphasise the resonance between sustainable farming and the teachings of the Vedas and other Hindu sacred texts. Palekar (n.d.), for example, stresses that the core of his approach to farming is the principle of non-violence, which he claims is derived from the Vedas, and contrasts to the ostensible violence of the Green Revolution. In a more esoteric vein, M. S. Deshpande (2003) presents an approach to farming based upon 'cosmic energies', as understood through Hindu scriptures. Though this use of religious expression in sustainable farming discourses may, in some instances, be an effective way of mobilising support, its more problematic features should not be glossed over. Some studies indicate a conservative tendency in sustainable agriculture movements in India. For example, Sharma (2006) reports that the watershed development project of Anna Hazare, who explicitly draws on right-wing Hindu('Hindutva') ideology, not only resulted in 'sustainable development' but also a tightening of the caste system and an exclusion of non-Hindus from the movement. Similarly, Nanda (2003) argues that certain critiques of the Green Revolution have a strong resonance with Hindu nationalism. She suggests that ideas relating to the 'harmony' and 'ecological balance' of traditional communities and farming systems that was disrupted by the Green Revolution serve to romanticise traditional hierarchies and obfuscate class, caste, and gender divisions within communities.

A Gramscian Approach: Sustainable Agriculture in the Context of Hegemony

My analysis of India's sustainable farming movement draws on the work of Antonio Gramsci, particularly his concept of hegemony. Gramsci (1971) refers to hegemony as an ongoing historical process by which dominant social groups consolidate social and economic systems that allow their interests to be served over long periods. In practice, they do this using combinations of persuasion, coercion, and deceit to win the support of subordinate groups for the systems they wish to establish. The ruling group may make concessions to subordinate groups to win this support—but only up to a certain point, as their core, long-term interests cannot be compromised. When a dominant group has secured ample support and cooperation from others, its view of the world becomes 'common sense' and its status in society can be said to be hegemonic: the ruling group then does not merely dominate; it *leads*. Although the use or threat of coercion remains present under conditions of hegemony, these are only a final resort—the reliance on consent far outweighs coercion.

Importantly, however, such a situation is never permanent, and no group's grip on hegemony is ever complete. Hegemony can be challenged, for example, when the dominant group fails to deliver promised benefits to subordinate groups or when new alliances or a new consciousness begins to develop amongst the general population, challenging the dominant view. This Gramsci (1971) refers to as a 'crisis of authority'—an event that may represent an opportunity to replace an old hegemonic regime with a new one. From a Gramscian perspective, the fate of particular social, economic, and political regimes thus does not rest solely on their technical merits but also on the outcomes of complex and contingent social processes to secure the active support of various social groups—particularly those with substantial social power and influence.

India's Green Revolution can be thought of as a hegemonic project. As I will explore in more detail in Chapter 2, at the time of its introduction, India's Green Revolution represented a convergence of national and transnational hegemonic interests, having been driven by the US state, US-based agri-business, 'big philanthropy', and certain class factions that were becoming dominant within the Indian state (Patel, 2013). The Indian industrial capitalist class favoured the Green Revolution as a strategy to boost food production, which it was believed would keep wage growth low. The Green Revolution succeeded as a *hegemonic* project, however, by enlisting the consent of medium to large commercial farmers, who were ascending to positions of social and political leadership within rural India. This class of farmers formed powerful movements in the 1970s, which demanded financial supports—such as subsidies and support prices—to ensure that the Green Revolution was a profitable venture for them. By conceding to a number of the demands of this class, the Indian state—and the class interests it represented—enlisted them as Green Revolution beneficiaries. They became crucial to ensuring sustained grassroots consent for this development agenda, which ultimately served the interests of industrial capital and (mostly foreign) agri-business. They allowed the Green Revolution to reach, if only temporarily, hegemonic status.

The development of agrarian crisis throughout India, however, represents a breaking down of this established hegemonic constellation. It is no longer possible to guarantee the active consent of large sections of the population for an agri-business-led development model. The rural development discourses that have been prominent since the 1960s no longer hold a hegemonic grip on the popular imagination—they continue to be in place only for a perceived lack of viable alternatives. This can be thought of as a Gramscian 'crisis of authority'. In this context, there is an opportunity for groups that had formerly been politically marginal to articulate and build consent for new development paths in the domain that Gramsci refers to as 'civil society'. The sustainable agriculture movement—at least in terms of its ideology—presents as one such attempt to forge a new path.

Various voices have emerged in India's civil society to articulate possibilities for more sustainable development and food systems. Particularly prominent have been NGOs, which are professionalised development organisations, formally independent of government departments but funded either by the state or donor organisations. NGOs typically introduce sustainable farming techniques in the form of development projects. Yet, in addition to NGOs, there have been several initiatives led and organised by members of rural communities, which focus on sharing knowledge, communal learning, and mutual empowerment. These initiatives may be termed 'social movements' or 'grassroots movements' (Dwyer, 2011). Between these two extremes, there are a diverse range of groups that might be termed as 'hybrid organisations' (Hasenfield and Gidron, 2005), which combine methods of organising from both NGOs and social movements. The case studies examined in this book belong to this 'hybrid' category, combining state or donor-sponsored projects with some form of grassroots mobilisation. Hybrid organisations prove highly useful in uncovering the intricacies of how hegemony operates in contemporary rural India. While NGOs have rightly been criticised for their connections to hegemonic institutions (e.g., Petras, 1999; Kamat, 2002), social movements are often assumed to be counter-hegemonic. Yet, in practice, the links between the two are often pervasive. Many NGO workers have personal backgrounds as activists and maintain ongoing connections to social movements (Ghosh, 2009). Furthermore, many NGOs draw on methods of organising that combine an organised professional component with a more organic, grassroots component (e.g., Appadurai, 2002). If we are to understand the intricacies of how hegemony operates, we need to explore the dynamic relations between these two forms of organisation as well as the ways in which hegemony is challenged, but also reproduced, through more 'grassroots' structures.

The sustainable agriculture organisations studied in this book frequently espouse ideologies that are counter to the hegemony of the Green Revolution and corporate food regime. They highlight the environmental and socio–economic damage caused by farmers' reliance on the chemical inputs of corporate agriculture. They openly critique corporate ownership of seed and promote control over natural resources by local communities in the interest of fostering the autonomy of small-holding farmers. Some of their initiatives go some way towards achieving these ends. Some organisations also have made efforts—with varying degrees of success—to challenge more localised expressions of hegemonic power—the dominance of upper-caste, large land-holding men over local politics, for example (especially the case presented in Chapter 5 of this book). Their work aims to build consent for new agricultural systems that empower small-holding farmers—and in some cases also more marginal groups, such as landless labourers—to take control over their farming systems. Their stated ideal is an alternative agricultural system for India, over which global agri-business has significantly less control and which uplifts poorer sections of rural communities and not only wealthier commercial farmers.

While these factors suggest that sustainable agriculture organisations have the *potential* to challenge the hegemony of agri-business over Indian agriculture, a core theme of this book is the ways in which they nonetheless serve to reproduce existing hegemonic relations. As I explore in greater detail in Chapter 3, sustainable agriculture organisations are embedded within a set of complex power relations that may undermine their counter-hegemonic impetus. While the authority of the state and ruling classes has been substantially undermined by current crisis conditions, sustainable agriculture organisations may play a role in reconstituting the state's authority. Indeed, a consistent theme in Gramsci's (1971) writing is how activities in civil society, rather than challenging state power, often serve to reinforce it, by direct or indirect means. Civil society is not a level playing field; not all people have equal opportunities to articulate (let alone implement) development alternatives. This is particularly the case in a country, such as India, where stark economic inequalities and multiple social hierarchies greatly affect people's access to the spheres in which public debate takes place (Chatterjee, 2004). Under certain conditions, sustainable agriculture organisations may subtly reinforce the current hegemonic configuration, not necessarily through the conscious conspiracy of elites but simply as an outcome of the uneven distribution of power and opportunities. This has major implications for the potential of sustainable agriculture to challenge the inequities of the corporate food regime.

In this book, I explore both the overt and covert ways that hegemonic forces shape sustainable agriculture organisations. On the more overt end of the spectrum, I explore the influence of the state and donor organisations, which exert considerable influence through their control over the funding supports for sustainable farming projects. In India, the state has legal mechanisms to cancel access to foreign funds for development projects if they are deemed to be 'political'. While the definition of 'political' in this context is highly vague, in practice, it has generally entailed the cancellation of funding for projects that are opposed to state interests (Jalali, 2008). Donor organisations may also restrict the counter-hegemonic potential of sustainable development projects. The dependency of most NGOs on external funding makes them more accountable to their donors and/or the state than to the communities with whom they work (Grey, Bebbington and Collison, 2006). As Riddell (2007) demonstrates, a very large proportion of NGOs receive their funding in the form of aid from rich countries, bilateral agencies, or multinational financial institutions—all of which have alignments with hegemonic interests. Of this funding, approximately three quarters is delivered to NGOs to implement projects that donors themselves have formulated, leaving NGOs very little autonomy. Since, in many cases, hegemonic powers are setting the agenda for development NGOs, it is unlikely that they will be directed towards efforts to undermine their authority by empowering the poor. It is for this reason that many scholars see NGOs as agents of neoliberal hegemony, rather than a progressive force (Petras, 1999; Sahoo, 2013). It has been observed that NGOs promoting changes that are harmonious with the neoliberal agenda, such as facilitating entrepreneurialism, market integration, and privatisation of commons, are more likely to receive funding than those that politicise development and give political voice to the poor (Ghosh, 2009; Patel, 2006; Gerber and Veutheys, 2010). Indeed, some studies indicate that NGOs, rather than empowering the rural poor, serve merely to pacify them and deflect public attention away from more radical, grassroots initiatives (Kamat, 2002).

On the more covert end of the spectrum, I explore the role of middle-class activists in obstructing the development of more genuinely counter-hegemonic movements for sustainable agriculture. The success of sustainable agriculture initiatives hinges largely on access to scientific expertise and skills in organisation and marketing, which have traditionally been the domain of the urban middle classes. The cases presented in this book show that, for this reason, urban middle-class activists and professionals often play prominent roles in leading and managing sustainable agriculture organisations. Yet, previous studies in India have cast doubt on the capacity of middle-class activists to adequately represent the rural poor, as they often rely on archaic and highly romantic images of rural life that do not reflect grassroots realities (Shah, 2010; Baviskar, 2004, 2005). Indeed, some studies have suggested that middle-class activism can be overtly antagonistic towards subordinate classes and social groups, taking a dismissive and condescending attitude towards their needs and aspirations and fostering relations of dependency (Anjaria, 2009; Baviskar *et al,* 2006; Fernandes and Heller, 2008). One of the unique contributions of this book is its clarification of the role of middle-class activists in sustainable agriculture organisations, both in providing crucial organisational support and in occluding the genuine political empowerment of those they claim to represent. It does not essentialise the Indian middle classes, which are highly heterogeneous, but explores how their involvement in the sustainable agriculture movement introduces a host of class interests that do not align neatly with those of the rural poor. For this reason, the counter-hegemonic potential of the sustainable agriculture movement—its capacity to challenge existing power relations—may be undermined in some instances by the continued prominence of urban middle-class activists as leaders of the movement.

A further factor that can subvert sustainable agriculture's counter-hegemonic potential is the influence of rural elites. One of the most problematic assumptions of sustainable agriculture advocates is that the farming methods that they promote can be beneficial and empowering to 'marginalised rural communities', and equating 'communities' with the village unit. This assumes that these villages are homogenous entities with homogenous interests, whereas, on the contrary, rural communities throughout most of India are highly divided along the lines of caste, religion, class, gender, and other factors. A major concern of the literature on rural development, therefore, is whether the rural elite are able to 'capture' the benefits of rural development projects, whilst presenting as though they are acting in the common good. Large and medium land-holding capitalist farmers have historically been a part of the hegemonic bloc in India, having been the chief beneficiaries of the Green Revolution. As such, they are more experienced in working with external agents (such as development workers) and speaking the technical language of development and may, thereby, use this capacity to ensure that the benefits of future development projects (including sustainable farming projects) flow on to them (Platteau, 2004). This book is highly mindful of this potential and explores the way in which, without due care, sustainable agriculture organisations may reproduce the hegemonic balance of power *within* communities.

While exploring these influences that may derail the counter-hegemonic potential of sustainable agriculture, I do not take the view that any of them are fixed or static. Indeed, one of the interesting features of the case studies presented in this book is the way that sustainable agriculture activists are often aware of these destabilising influences and attempt to develop strategies to overcome them. Here, the point is to develop a dynamic perspective on the connections between hegemony, sustainability, and rural development, highlighting the barriers to truly reworking the status quo, and whether there might be effective ways to overcome these barriers.

In short, I take up the question: what is the role of sustainable agriculture organisations in sustaining and/or challenging relations of hegemony in rural India? Do they facilitate the rural poor to challenge existing hegemonic relations and assert alternatives that better meet their needs? Or do they merely reinforce the status quo? To express this in another way, can their leading activists be considered exemplars of Gramsci's 'organic intellectuals', who give voice to grassroots perspectives, build unity, and educate, or are they rather 'traditional intellectuals', who reinforce existing imbalances of power? To address these overarching questions, two interrelated issues are explored. The first issue pertains to the organisational structure and strategy of organisations and whether they provide suitable spaces for the participation and empowerment of the rural poor. The second issue relates to the broader power structures and networks of social relations in which these organisations are embedded. Implementing sustainable agriculture projects requires activists to forge relations with various parties, to source funds, link to experts, and build ties with the communities that they are hoping to change. I explore how these relations affect organisations' methods of operating and the outcomes of their work. If organisations are found to have stronger relations with donor organisations, corporations, rural elites, or the state than they do with the rural poor whom they claim to serve, then it is doubtful that they can challenge the current hegemony in any meaningful way. In this way, I interrogate whether sustainable agriculture organisations reflect and/or challenge existing forms of social power.

Case Studies and Methodological Framework

Addressing the questions raised by this book required detailed attention to largely qualitative phenomena. It was necessary to explore the structure and class composition of organisations, the way in which activists conceptualise contemporary conditions in rural India, the benefits that they ascribed to the farming models that they promote and their strategies for promoting these models. It was also necessary to attend to the social relations that organisations formed to achieve their goals, their processes of engaging with rural communities, and the impact of these relations and processes on outcomes. To investigate these factors, I built a rich data set through use of interviews (interviewing a total of

109 people), participant observation, and primary and secondary materials. The data was collected first in the context of my doctoral research, between November 2009 and April 2010. This has been supplemented by follow-up research visits between February 2014 and January 2015.

In the course of this book, I present three case studies of sustainable agriculture organisations operating in India today to illustrate in close detail the ways in which their activities are structured by and challenge relations of hegemony. I adopted a case-study approach, primarily because understanding issues of social power and hegemony requires engagement with factors that vary regionally, including the structure of civil society, agrarian power structures, relations between rural and urban societies, and relations between civil society and the state. To do justice to these complex issues demands close attention to local and regional contextual factors. A case-study approach allows the collection of rich, detailed, and variegated data that provides clear illustrations of social processes within a regional context. It allows for useful, comprehensive data to be collected in instances when it would not be possible or appropriate to perform a complete survey of the field of interest (Yin, 2009). To attempt a complete survey of all organisations involved in promoting sustainable agriculture in India would not only be beyond the scope of this book but would also risk losing focus. In-depth data on a small number of carefully selected case studies better illustrates the social processes that are the main focus of the book.

The regions in which the three case studies are based reflect the diversity of agrarian conditions in contemporary India, in terms of their level of commercialisation and use of Green Revolution technologies. The first case study is based in Punjab, where a resource intensive, commercial approach to agriculture has been in place since the introduction of the Green Revolution in the 1960s. The second case is based in Tamil Nadu, which has a more complex history with commercial agriculture. The Green Revolution came to some districts of Tamil Nadu, specifically those with established irrigation facilities, but not others. Furthermore, due to the small size of average land holdings, there was less potential for the Green Revolution to be profitable for farmers in the state. The third case study is based in Uttarakhand, which never really had a Green Revolution to speak of and whose farming is mostly oriented towards local subsistence requirements, rather than the market. Each of these regions experiences agrarian crisis in a unique way. The regions that have followed a Green Revolution path of agrarian development tend to be more affected by ecological issues, the rising cost of chemical inputs, and indebtedness. In regions that did not follow this development path, farmers tend to be poorer and are more sensitive to deflation of food prices and inflation of the cost of living. Furthermore, agriculture in these areas tends to be rainfed and is therefore more vulnerable to the impact of climate change. By examining organisations that are operating in such diverse conditions, I draw attention to the ways in which sustainable agriculture is being positioned as a solution to a range of contemporary issues but also the unique challenges of operating in regions with divergent class structures and hegemonic configurations.

The organisations presented in this book can all be described as hybrid organisations, combining, to various extents, funded development projects with aspects of grassroots social mobilisation. As such, they perfectly embody some of the contradictions of the sustainable agriculture movement at large: they engage with both the mainstream development establishment on the one hand (more often in alignment with existing hegemonic relations) and rural populations and the broader sustainable farming movement on the other (which offer potential for counter-hegemonic articulations). The methods that each of the three organisations use to negotiate these contradictions help to illuminate the competing pressures on the movement at large.

Each case study provides an illustration of the different ways in which these tensions are negotiated. The first case study, the Kheti Virasat Mission (KVM) of Punjab, adopts a largely rural populist approach, presenting sustainable farming as a means of building health, harmony, and resilience for communities, whilst in practice often prioritising the perspectives of relatively privileged land-holding farmers and the urban middle class. The second case study, The Tamil Nadu Organic Farmers' Movement (TOFarM), made clear and conscious efforts to overcome any elite bias by facilitating the participation of marginalised sections of communities in their movement for sustainable change. Nonetheless, the status of its leading figures as urban outsiders and their dependency on the development establishment created challenges to fostering a movement that truly reflected these marginalised perspectives, in some ways perpetuating hegemonic relations. The final case study, the Beej Bachao Andolan (BBA) of Uttarakhand, appears to overcome some of these limitations, as it is an initiative to defend subsistence agriculture created and led by members of rural communities, offering the possibility of a grassroots articulation of resistance to the hegemony of the corporate food regime. Nonetheless, its gradual integration into circuits of middle-class activism may have subverted this counter-hegemonic potential.

These organisations were first identified through internet searches. Their prominence and appropriateness for the research project were verified through contacts who had been working in reputed national and international NGOs for several years. These contacts introduced me to the leaders of the organisations through phone or email. With further correspondence, their interest in participating in the research was established and plans were made for my initial research visit.

For the most part, the anonymity of research participants has been preserved. However, for the leading figures of the organisations studied—namely, Umendra

Dutt, M. Revathi, Vijay Jardhari, Dhoom Singh Negi, Sudeshi Devi, and Biju Negi—this was seen to be redundant. Each of them are public figures, and anyone with a reasonable knowledge of the rural development scene in their respective states would be able to guess their identity, even if they were provided with pseudonyms. In some cases, these leaders also did not want to be anonymised. I thus acquired their written consent to use their names in publications. All other research participants who are mentioned in this book have been given pseudonyms.

For each of the three cases, I collected diverse forms of data using several methods. These methods were drawn upon selectively, as was most appropriate for the particular case. Qualitative interviews provided the most substantive data source. They provided insights into the thoughts and experiences of those involved in the organisations and provided crucial details on organisations' histories, objectives, strategies, and the social relations they forged to achieve their goals. Broadly speaking, interviewees fell into three main categories: activists, professionals, and members of target communities (whom I refer to as 'villagers' below, for the sake of brevity). This distinction has been made as each of these groups provided unique sources of information. Table 1 shows the number of interviews conducted with each of these categories of participants, for each case study.

Table 1: Interviews Conducted in Each Location

	Activists	*Villagers*	*Professionals*	*Total*
Punjab	13	14	21	48
Tamil Nadu	10	19	1	30
Uttarakhand	12	15	4	31
Total	35	42	26	109

'Activists' refers to those who have devoted their time and labour to the organisations in a sustained manner and include project workers, volunteers, and campaigners. They are involved in activities, such as organising communities, arranging meetings and workshops, lobbying government, promoting sustainable techniques amongst farmers, teaching new farming techniques, and so on. In interviews, they were asked a range of questions, which varied depending on the interviewee and the work they were engaged in. Amongst other things, they were asked about the history of their organisations, their views on crucial issues, such as the Green Revolution and the benefits of sustainable farming, their relations with donors, the state and activist communities, their approach to engaging with different sections of rural communities, and the challenges they had faced. These interviews were particularly useful in clarifying organisations' discourses, strategies, social relations, and histories.

'Villagers' were those who live in rural communities and were the targets of the organisations' interventions. In this category, I include farmers, labourers, women, and other members of rural communities whom organisations tried to involve in some capacity. Amongst other things, they were asked about the state of agriculture in their village, how they came into contact with the organisations, whether they found their message compelling, their reasons for being attracted to sustainable agricultural practices (or not), and their experiences in adopting these practices (if they had used them). Their interviews provided insight into the reasons that farmers were adopting new agricultural techniques, their reluctance to make further changes, and their experiences with the organisations in question. Importantly, this allowed the claims of activists regarding their work and impact to be cross-checked against those with first-hand experience and less incentive to present ideologically skewed interpretations. Most of these interviews were conducted with those who were in some way involved in transitioning to sustainable agriculture. Some, however, were conducted with those who were interested in making the change but nonetheless reluctant—their interviews providing key insights into reasons certain sections of rural communities are reluctant about adopting these methods or aligning with the organisations in question.

The category of 'professionals' refers to supporters, such as academics, journalists, medical practitioners, lawyers, teachers, and bureaucrats. These participants were less intimately connected with the inner workings of the organisation than activists. Instead, they drew on their own professional expertise to provide various forms of support. They were asked about their views on and experiences with the organisations and the issues they were engaging with and the nature of their support. Their input was useful in providing outsiders' perspectives on the organisations and showing how organisations had an impact outside of their immediate spheres of influence.

It must be noted that the boundaries separating these categories of interviewees were often blurred. In Uttarakhand, for example, many of the most important activists are also farmers. Furthermore, often the people who provided 'professional' support were also doing important organisational work and as such could also be categorised as 'activists'. The figures in Table 1 therefore provide only a general outline of the composition of interviewees. Interviewees were selected on the basis of their salience to understanding the everyday operations of the organisations and their impacts within the communities. Consequently, the number of interviewees from each of the categories listed above differed between the three case studies. In particular, a higher number of 'professionals' were interviewed in Punjab, since the organisation studied in Punjab was more reliant upon professional support and the views and experiences of these professionals were more relevant to understanding the organisation. The recruitment of interviewees occurred initially through the organisations' leading activists, who would provide the necessary contacts required. Though this was convenient in the early phases of research, it soon became apparent that activists would act as 'gatekeepers', only providing access to interviewees who would provide positive perspectives on their initiatives. Thus, once I became more established in each field site, I would seek out independent translators for assistance and venture out into the organisations' target villages on my own to speak with organisations' intended 'beneficiaries'. The one exception to this was the Uttarakhand case study, where I was embedded in the village from the outset and had no challenges in recruiting local participants for interviews. Where possible, I also made efforts to reach out to activists and professional supporters independent of organisations' leadership, including those who might be critical of aspects of the organisations, to develop a more balanced perspective.

Interviews varied in their length and level of formality. Some were relatively unplanned interactions without any formal structure; others followed a more formal course. The vast majority of interviews, however, were semi-structured. There were certain interview questions that were prioritised for each case study. Individual interviews, however, did not follow a precise schedule. Adopting a rigid structure to interviewing can render the researcher insensitive to the contingencies of the case setting (Devereux and Hoddinott, 1993). Questions were prioritised and adapted according to the participant and the information it was anticipated they would be able to provide. Following the semi-structured approach, a fixed set of priority questions were taken into each interview, but interviewees were given some scope to influence the direction the interview would take. When an unanticipated issue of interest arose, interviewees were asked for clarification and further details.

The second major source of data was first-hand participant observation. As Denscombe (2007: 206–225) describes, this method provides a number of benefits when combined with other methods, enriching data and increasing its reliability. It enabled me to collect primary data on participants' activities, reducing reliance on information collected second hand through interviews and secondary materials. Furthermore, unlike in an interview, simply observing occurrences without providing input regarding my own views or expectations, reduced the observer effect which may occur when participants guide their responses to align with the perceived views of the researcher. By taking a more passive role and observing, there was greater potential for unanticipated issues to reveal themselves. This was particularly the case when it came to issues regarding caste and class discrimination within organisations, which were more often revealed in candid settings. Finally, participant observation allowed me to develop an understanding of the everyday activities that groups are involved in and the more quotidian issues that they confront, which may not have arisen in the context of an interview.

Participant observation was conducted in several contexts. At the most basic level, it involved taking note of the day-to-day activities of people involved in the organisations. During my fieldwork, wherever possible, I stayed in the homes of participants of the movement, which allowed me to observe how their work in the organisation structured their daily lives and perspectives. This also created possibilities for more candid, frank interactions. For the case studies in Punjab and Tamil Nadu, I was able to observe some of the background work that goes into building organisations and planning for public events, such as networking, promotion and fundraising. Participant observation was also conducted with farmers. A particularly common form of interaction was for farmers to take me on a tour of their farm, showing the kinds of changes they had made since becoming involved in sustainable agriculture. This provided opportunities for casual discussions regarding farmers' experiences of success and failure, their perceptions of the kinds of support they received and so on. Another important instance of participant observation was my participation in public activities, such as workshops, lectures, and public meetings. Observing these activities allowed me to develop an understanding of how the organisations project their message to the outside world, interact with various audiences and network.

The nature of the data collected via participant observation was unique. It revealed the intimate details of the work being done and the everyday challenges involved. As Devereux and Hoddinott (1993) note, being embedded within an organisation allows one to gain a greater appreciation of the cultural context in which activities occur. This was particularly the case in Uttarakhand, where during the course of fieldwork I stayed in a village and lived with a farming family who were themselves involved in the organisation under investigation. I was able to gain an understanding of the cultural context that gives meaning to the more public acts of the organisation. It also allowed greater rapport and trust to develop between myself and participants, facilitating the communication of more detailed and in-depth data. Living in the village provided more opportunities to interact with farmers and gain an appreciation of the issues they were facing and the ways in which activists engage with those issues.

A third data source was primary written materials produced by the organisations themselves. For the most part, these were promotional materials, instruction manuals for farming techniques, and documents stating the organisations' recommendations to policy-makers. Often, they revealed information about the philosophy of the organisations and the manner in which they project themselves to other social groups. They also reflected the impact activists hoped to have on farmers and public debate. Typically, they were composed by members of the organisations themselves or by allied groups, such as larger NGOs or donor organisations. As a general rule, the statements made in such materials were less reliable than those made in interviews. They were more likely to reflect what activists wanted the outside world to believe about their organisations and the farming systems they promote, rather than the organisations' actual activities and activists' real perspectives. As Edelman (2009) notes, these 'official narratives' are nonetheless important, as they reveal how organisations engage with contemporary issues at the level of representation. Furthermore, it may be of interest to analyse the gap between public image and actual practice, which requires promotional materials to be analysed in close conjunction with data collected through other methods. This gap is of particular interest in this book. As previous studies have shown, distorted representations of certain issues by activists can have problematic consequences for the claimed 'beneficiaries' of such action (Baviskar, 2005; Shah, 2010). The gap between activists' representations of sustainable agriculture (in both public materials and in interviews) and the experiences of target communities were thus a major focus of analysis.

The final data source was secondary written sources, which were used to provide context and meaning to the data, as well as a means of cross-checking and validating claims made by those within the organisations. They were of three types. First, academic articles were used to give a broad understanding of regional context. Factors such as agrarian histories, social structure, and contemporary politico–economic, and ecological issues were all considered. Secondly, local English language news publications were followed during the research period. This helped to develop an understanding of the most pressing issues in the regions at the time and the extent to which the organisations were engaging with those issues. On a few occasions, the organisations themselves featured within the local newspapers, which allowed me to gain some perception of how they were being represented in the media to the general public. Thirdly, on a few occasions, I was able to collect materials written about the organisations in question by third parties. These third parties included other NGOs, activist networks, and special interest publications, such as magazines.

Taken together, these various data sources provided comprehensive and balanced perspectives on each of the three organisations. It was possible to gain an in-depth understanding of their stated objectives and public narratives, whilst also cross-checking these against the experiences of the rural communities and external agencies with whom they interacted. By spending extended periods of time with each of these organisations, I was able to gain a reliable perspective on their impact. Furthermore, by interviewing and observing a diverse range of relevant participants, I was able to decode the ways in which organisations' position within social relations (particularly their relations with hegemonic forces) served to guide and structure their activities in various ways that had implications for how the concept of 'sustainable agriculture' unfolds on the ground.

Synopsis

In brief, throughout this book, I explore the paradoxical position of sustainable agriculture organisations in relation to hegemony through reference to detailed case studies. I consider the counter-hegemonic potential of sustainable agriculture but also delve critically into the ways in which sustainable agriculture activists may become complicit in serving hegemonic interests and reproducing hegemonic relations. Leaving some space for hope, however, I also consider the agency of activists in recognising the constraints imposed upon them by their position within social structures and their creative attempts to overcome these constraints. The book opens discussion on possible ways forward, such that development workers, activists, and scholars alike may employ sustainable agriculture in ways that are more genuinely empowering for marginalised members of rural communities and pose a more meaningful challenge to the status quo.

In Chapter 2, I develop a Gramscian perspective on the history of rural India. I explain Gramsci's concepts of hegemony, civil society, the state, and the 'crisis of authority' and apply these to the specific conditions in rural India from Independence to the present day. The chapter provides detail on the kinds of development models that have been pursued in India during this time, some of the issues associated with these models and, crucially, how these models reflected and reproduced hegemonic interests. In this chapter, I advance the argument that the current scenario in rural India may be interpreted as a Gramscian 'crisis of authority', in which the leadership of currently hegemonic groups and ideas are breaking down. If this is the case, it suggests that there is an opening for the development of alternative approaches, of which sustainable agriculture forms one example.

Chapter 3 develops a conceptual framework to assess the position of sustainable agriculture organisations within structures of hegemony. It begins by outlining some of the resource needs of these organisations, highlighting the fact that these needs often require external support, including skills, social connections, and funds. This support, at present, is typically drawn from urban middle-class development workers, the state, and donor organisations (particularly transnational donors). Meanwhile, when working in rural communities, sustainable agriculture activists must make choices about which sections of communities they wish to work with. I consider that the strategic concerns at play may lead organisations to align with either rural elites or the rural poor. I describe sustainable agriculture organisations as dynamic entities, tenuously positioned between the often-competing interests of the state, donors, middle-class activists, the rural elite, and the rural poor. I argue that the way in which organisations manage their relations with these various groups has major implications for whether their projects challenge or serve hegemonic interests.

In Chapter 4, I introduce the first case study, the Kheti Virasat Mission (KVM). This initiative has been working to promote what its leading activists call 'natural farming' in the state of Punjab, the heartland of India's Green Revolution. Given the pervasive use of chemicals and intensive agricultural practices, introducing natural alternatives in this state is certainly an uphill battle. This is particularly the case, as the Punjab state government has historically been strongly committed to promoting Green Revolution technologies amongst farmers. KVM's leaders have exhibited a strong sense of distrust towards the state and donor organisations (particularly transnational donors), recognising these to be key forces behind the Green Revolution and corporate approach to farming to which they are opposed. As such, rather than taking on the approach of a typical NGO of receiving funds from donors or the state and using these to implement projects for sustainable development, KVM has instead styled itself as a 'people's movement', working in an organic way with the local population to promote change through all layers of society. The paradox that I explore, however, is that this does not allow KVM to completely escape from hegemonic influence, for two key reasons. First, styling itself as a 'people's movement' with an activist disposition has involved large numbers of middle-class activists within the organisation. Being highly dependent on the urban middle class for media coverage and the receipt of regular, small-scale private donations, KVM has tended to focus more on issues that bolster their righteous image amongst middle-class audiences, often to the neglect of issues of more direct relevance to farming communities. This highlights the ways in which the hegemony of the urban middle class over ideological matters in India encroaches on sustainable agriculture organisations and limits their scope. Secondly, KVM's claim to be a 'people's movement' representing all of Punjabi society degenerates into a kind of rural populism on the ground. While claiming to represent the marginalised, their projects are of benefit predominantly to land-holding men. For women and the landless, KVM's methods of engagement (or lack thereof) tend to reproduce the status quo or even further entrench disadvantage and marginalisation. The implication of populism for rural development strategies is considered in detail.

The organisation presented in Chapter 5 took an approach that was, in many ways, more mindful of the risks of external influence and elite dominance than KVM. The case study is the Tamil Nadu Organic Farmers' Movement (TOFarM), an initiative that became well known for its work in tsunami restoration in the Cauvery Delta of Tamil Nadu. The leader of this initiative, M. Revathi, followed an approach that has become popular in the rural development community of avoiding the assumption of a 'leadership' role within her organisation and instead positioning herself as a 'facilitator'. Her objective was merely to facilitate rural communities to drive their own movement for sustainable farming, by creating spaces in which they could devise plans to use ecological farming techniques to meet their own immediate needs and aspirations, rather than imposing plans on them from the top down. Furthermore, to avoid elite dominance, Revathi devised strategies to deliberately involve marginalised sections of communities, including women, landless labourers, and Dalits (the caste at the lowest level of the Hindu hierarchy). In forming a movement, rural people were encouraged not only to disseminate ecological farming techniques for their own economic empowerment but also to collectively 'raise their voice' and make political demands to improve their conditions. As such, the initiative, at least theoretically, had great potential to alter power relations and challenge the existing hegemony in rural India. Nonetheless, despite good intentions, the initiative was still restricted by the power relations in which it was enmeshed. I explore how the development establishment, local elites, the state government, and other local and non-local NGOs all attempted to influence the initiative in ways that derailed its core objective of promoting sustainability in a way that empowers the rural poor.

Chapter 6 presents a case study qualitatively different to the other two, the Beej Bachao Andolan (Save Seeds Movement, BBA) of Tehri-Garhwal, Uttarakhand. Unlike KVM and TOFarM, whose leaders were from urban, non-agricultural backgrounds, BBA is organised and led by members of poor rural communities. The movement began as a grassroots initiative of villagers who, inspired by the famous 'Chipko' forest conservation movement, amassed a diverse variety of seeds from various parts of the Uttarkhand Himalaya, to make them available in collective village seed banks. The initiative was intended to preserve local agro-biodiversity and to encourage farmers to reject Green Revolution monocultures and the commercialisation of local farming practices, which had disrupted the subsistence economy and left farmers vulnerable to market and environmental shocks. Having a diverse variety of seeds available also helped to bolster the biodiverse subsistence farming traditionally practised in the region and leave it stronger in the face of climate change. As a movement coordinated by members of rural communities in a region with a relatively flat agrarian structure, BBA seems on the surface to be an ideal example of how sustainable farming initiatives can be oriented towards the needs of the rural poor and challenge the hegemony of the corporate food regime. While this is largely true, the case study also reveals how even an initiative led by the rural poor can come under the influence of other agents who can redirect its counter-hegemonic agenda. While BBA rejected relations with the state and donors, it has formed extensive links with middle-class activists spread throughout India and, indeed, the world. Being embedded in these networks has led BBA to orient its message more towards the interests of middle-class audiences around the world and, in recent years, this has led to the neglect of more salient local issues, such as climate change and rural out-migration. The case study shows subtle workings of hegemony, in which, even in the absence of any conscious conspiracy, the balance of social power can defuse the counter-hegemonic potential of genuinely empowering movements.

In Chapter 7, I draw out the theoretical implications of the research and attempt to offer some direction for pathways forward. I argue that the case studies demonstrate the diverse ways in which sustainable agriculture organisations become ensnared in hegemonic relations and, thereby, often serve to reinforce the status quo, rather than challenge it. Particularly problematic, I argue, is the fact that the leading activists of sustainable agriculture initiatives tend to adopt 'insular' dispositions, taking input more from fellow activists and a limited number of land-holding farmers, rather than from the rural poor. This limits the potential of sustainable agricultural interventions to challenge existing power relations and provide empowering solutions to the agrarian crisis. I draw out two separate sets of implications from these key findings, intended for two separate audiences. First, for those working in the development community, I emphasise the importance of development workers and activists paying close attention to local and non-local power dynamics and limiting their interventions to instances in which broad sections of the community will derive substantial economic empowerment. Ultimately, I argue that it may be more productive for activists to move beyond an exclusive focus on sustainable agriculture, as a pre-formulated project, and instead attempt to augment and expand upon existing patterns of resistance amongst the rural poor. Second, for those interested in counter-hegemonic resistance to the corporate food regime, I argue that the fixation on the concepts of 'food sovereignty' and agro-ecology in both activist and scholarly literature is perhaps not suited to the Indian context. For the majority of rural communities in India, more meaningful forms of resistance to hegemonic relations may present in less obvious forms: particularly in their leverage of progressive policy mechanisms to improve their position in both the local rural economy and in global food-supply chains.

These conclusions, which are largely critical in nature, should be read in the spirit in which they are intended. The book proceeds with an interest in developing sustainable responses to agrarian crisis that empower the rural poor to meet their needs in a manner that reflects their own perspectives and interests, rather than those of elites. The critiques I present aim to provoke inquiry into ways that sustainable agriculture organisations could be more effective and inclusive, and are therefore offered, to an extent, as an expression of solidarity, acknowledging a mutual commitment to a common cause (Scheper-Hughes, 1995). I acknowledge, from the outset, that such critique is much easier from the academic armchair. As Edelman (2009: 256) notes, activists in developing countries often operate in a 'permanent crisis-response mode'. I have endeavoured to be sensitive to the tight working conditions that activists live under and, indeed, much of my critique throughout this book is directed more at this constrained social context than at individual activists or organisations.

2

India's Agrarian Crisis

A Gramscian View

Low external input sustainable agriculture is often presented as a solution to an 'agrarian crisis' said to affect much of rural India. Broadly speaking, this crisis represents a breakdown in the functioning of rural social, economic, and ecological systems, such that the agrarian sector has a diminished capacity to provide sustainable livelihood options for the country's largely poor rural majority. This chapter puts the idea of an 'agrarian crisis' into historical perspective. Throughout the chapter, I adopt a Gramscian theoretical framework, highlighting the ways in which hegemonic groups have influenced India's rural development trajectory since Independence. This entails a detailed analysis of the classes and class fractions that have benefitted from dominant approaches to rural development and the ways in which they have ensured that these benefits continue to flow in their direction. Furthermore, it explores how developments in civil society have aided these hegemonic groups in establishing popular consent for a largely inequitable food system. Ultimately, however, I argue that this consent has now broken down, such that the agrarian crisis has now led to a 'crisis of authority' in which the credibility of hegemonic groups and ideologies in rural India can no longer be sustained.

Discussions of India's agrarian crisis tend to oscillate between despair and utopianism. The current scenario is presented as bleak, deeply entrenched, and degenerative, while the proposed solutions lack the clarity of a precise political agenda. It is in this sense that a Gramscian perspective can make a unique contribution. Gramsci encourages an analysis of social processes already in motion that may lead to the development of alternative social, economic, and political forms (imperfect as these may be). In arguing that contemporary rural India can be analysed not only as an 'agrarian crisis' but also as a Gramscian 'crisis of authority', I suggest that the existing hegemonic structures that have maintained consent for particular models of agrarian development are no longer effective. As the old hegemony loses its grip on the popular imagination, new ways of thinking about the future can emerge. It is thus possible for new alternatives to be articulated (low external input sustainable farming being just one), as formerly subaltern groups begin to build alliances with those disaffected by the current approach, establishing new bases of power. This process takes place within the domain Gramsci described as 'civil society'.

The chapter begins with a brief introduction to Gramscian theory, placing particular emphasis on the important concepts of civil society, hegemony, and the state, which informs much of the book's theoretical analysis. Adopting this theoretical framework, I then outline the factors that led to the modernisation of Indian agriculture under the banner of the 'Green Revolution' in the 1960s and 1970s, revealing how this massive development intervention was driven by and ultimately reinforced the power of hegemonic groups. It then turns to explore the (largely negative) long-term social, economic, political, and ecological impacts of the Green Revolution and explains how these impacts have undermined the legitimacy of the Green Revolution and the social groups who traditionally had promoted it. I then explore the impact of the liberalisation of India's economy in the early 1990s, considering its reconfiguration of hegemony in rural India and its impact on the agrarian sector. I conclude by explaining why the current situation is regarded as an agrarian crisis, presenting my argument for conceptualising this as a Gramscian 'crisis of authority', and exploring a few of the dominant responses to this crisis.

Civil Society, Hegemony, and Crises of Authority

The concept of civil society has been an important component of Western social and political theory, having been used extensively by Enlightenment philosophers, such as Hobbes and Locke (Fontana, 2006). Since the 1980s, the concept has re-emerged as a central focus of academic debate in the English-speaking world, as scholars attempted to explain the development of popular movements against dictatorships (Cohen and Arato, 1994). In this context, the term has been equated with the 'public sphere': a space independent of the state that balances state power and keeps it accountable (McLaverty, 2002). From this perspective, the development of a strong and independent civil society is crucial in preventing tendencies towards dictatorship and the arbitrary exercise of state power. This approach, which defines civil society in terms of its non-state and anti-authoritarian characteristics, has become popular in contemporary readings of civil society in academic and activist circles.

This view of civil society has been critiqued by Gramscian scholars. Buttigieg (2005) argues that such an approach, which paints civil society as simply an 'an ensemble of popular progressive oppositional movements not formally or necessarily affiliated with a particular political party' (Buttigieg, 2005: 34), has little value to political theory or strategy. These oppositional movements do not make up the entirety of civil society, and to focus exclusively on them would be to neglect civil society's sociological and political significance. For Gramsci (1971), civil society is not only the space in which opposition is formed and expressed; it is also the ground on which consent for state activities develops. The state depends on civil society to operate with the consent of the general public and will at times intervene in civil society to establish such consent. Therefore, the state and civil society should not be considered two distinct spheres: each presupposes and permeates the other, such that, from a Gramscian perspective, they can only be separated for 'methodological' purposes (Buttigieg, 2005: 39).

This closely relates to Gramsci's theory of hegemony. Indeed, as Fontana (2006: 55) puts it, from a Gramscian perspective, civil society is 'the locus of hegemony'. Hegemony can be best explained as an historical process by which a particular social group comes to occupy a ruling position in society and how they maintain that position once they have gained it. Gramsci (1971: 180–181) emphasises that any group that aspires to leadership status must develop modes of cooperation with other social groups with whom it has common interests. In addition to this, however, to be truly 'hegemonic', the group must learn to guide the interests of subordinate groups—whom Gramsci refers to as 'subaltern'—such that they become harmonised with its own interests. Hegemony thus refers to this process by which particular groups come to take on a leadership role, partly through the use of coercion or the threat of coercion, but, to a greater extent, through persuasion and consent-building. While the central objective of the hegemonic group may be to gain control of the state and establish a legal apparatus through which its interests can be continually served, it is only by way of enlisting the consent of other groups in the domain of civil society that this becomes possible.

Once the hegemonic group has state power, they are able to use the state apparatus to manipulate and intervene in civil society, 'civilising' the population such that their interests (real and perceived) align with dominant interests. Often this influence is subtle and indirect. The techniques used by the state to indirectly build consent have been the subject of much analysis by Gramscian scholars (Lears, 1985). Kamat's (2002) analysis of civil society groups in Maharashtra, India, provides a clear example of such 'indirect' influence. Kamat shows how, through sponsoring and elevating the position of civil society groups whose perspective is non-threatening to hegemonic interests, the state scatters, neutralises, and marginalises groups that are in fundamental opposition to the hegemonic perspective. Non-threatening groups thereby command greater influence over public opinion—they encourage forms of activity that are suitable to hegemonic interests. When these kinds of interventions into civil society are successful, the two entities, the state and civil society, begin to coalesce as an organic whole—part of an 'historic bloc' that protects certain interests and promotes a particular form of economy, society, and way of life.[3]

In concrete historical situations, however, the state and civil society always fall short of total organic unity. While much of the emphasis in Gramsci's writings is on how political society must build a framework of consent in civil society to operate effectively, it is also apparent that at times this framework is stronger than others. Indeed, Gramsci (1971) clearly recognises that, at certain historical periods, the state and civil society are fundamentally disconnected. This is most apparent in periods of 'organic crisis' or 'crises of authority' in which dominant groups are only able to resort to coercion to maintain their rule—they are unable to play a leadership role in building consent through civil society. The situations in which this may take place are diverse, yet Gramsci (1971: 210) suggests two common causes:

> [E]ither ... the ruling class has failed in some major political undertaking for which it has requested, or forcibly extracted the consent of the broad masses (war, for example), or because huge masses (especially of peasants and petit-bourgeois intellectuals) have passed suddenly from a state of political passivity to a certain activity ...

In a capitalist context, economic developments are also crucial in this process. The institutions that the hegemonic group develops to maintain consent for its rule are ultimately unable to contain the disruptive effects of the uneven development of capital (Martin, 1997; Saull, 2012). In a period of crisis, the state lacks the means to respond to these effects through consensus (being constrained by the existing institutional framework) and hegemony is weakened. In the resulting power vacuum, formerly subaltern groups, who previously lacked political representation, are able to emerge in new leadership positions within civil society. During such times, disaffected groups are well positioned to begin a process of forming alliances and gaining the consent of other groups, leading to the formation of new hegemonic projects. In this way, civil society, while being the domain in which ruling groups legitimise their power, can also be the domain in which alternative constellations of

[3] This attention to the subtle and indirect ways that the state is involved in shaping the lives and consciousness of subaltern groups is one of the key reasons that Gramscian analysis has proven useful in understanding the social and political dynamics of contemporary India. While earlier scholarship, such as that emerging from the 'subaltern studies' school had tended to see the state and subalterns as fundamentally disconnected in India (e.g., Guha, 1998), recent scholarship has taken more direct inspiration from Gramsci in investigating the complex ways in which the state and subaltern are intertwined through relations of both consent and coercion (Nilsen, 2015; Nilsen and Roy, 2015).

power are formulated. Indeed, as subaltern groups come to articulate themselves in ways that challenge hegemonic worldviews, civil society can become the birthplace for a new socio–political ethic.

Hegemony in Rural India

Industrial development has been the chief hegemonic project of most postcolonial states, including India (Escobar, 1998, 1992; Sinha, 2003). Often emerging from the colonial experience with a crippled industrial sector, postcolonial states have set themselves the task of rapidly shifting from an agrarian to an industrial economy, to become competitive with the more economically-advanced capitalist nations. From an economic standpoint, this task is challenging, as postcolonial states generally lack the savings to finance industrialisation. In this context, the agrarian sector is looked upon as a potential source of cheap, exploitable resources, in the form of capital, labour, and food, to fuel the development of industry (Varshney, 1995: 14–20). This is mostly done either through taxation of agrarian surpluses or adjusting the terms of trade in favour of cities and to the detriment of the countryside (Byres, 1991). The problem was first formulated by Preobrazhensky (1965), in the Soviet Union, under the heading of 'agrarian question', and later by Byres (1991), who described it as 'agrarian transition'. Byres describes how agrarian transition occurs in all major economies as they industrialise, but the processes by which this transition occurs vary as a function of the local balance of class forces. This suggests that the challenge of facilitating agrarian transition is not only technical but also *hegemonic*. It is highly difficult for the state to retain the consent of agrarian populations for such a large, often inequitable transfer of resources. Many states resorted to coercion—most notably the Soviet Union, in which peasants were forced onto state-managed, collectivised farms and potentially resistant landlord classes were arrested, killed, or deported (Conquest, 1986). Consent is more challenging since the reaction of agrarian classes to these policies has generally been antagonistic. One approach has been to ensure that certain agrarian classes receive benefits from agrarian reforms, thereby dividing the countryside and preventing the formation of a rural bloc that can coherently articulate dissent (Bates, 2005).

India was no exception to this historical tendency. After Independence, Prime Minister Jawaharlal Nehru sought to rapidly industrialise the national economy. Yet, conditions for achieving this were challenging. Capital was in short supply. British colonialism had operated largely through coercion and had suppressed industrial production in India's cities, while empowering a class of rural intermediaries in the countryside (known across much of India as *zamindars*), who oversaw an extractive economy that was often semi-feudal (Guha, 1998). This made the position of the industrial capitalist class weak. Economically, this class was dependent on non-capitalist sources of accumulation, mostly in the countryside, to provide the capital to fuel its growth (Sanyal, 2007). At the time, there was broad political consensus within Nehru's ruling Congress Party that the best way for this process to occur was to keep food prices low by increasing agricultural production—as low food prices would put downward pressure on industrial wages (Varshney, 1995). Yet, any change to agrarian relations to increase production was difficult, as the socio–political field was dominated by rural elite classes. The industrial capitalist class, though having the support of the leadership of the Congress Party, did not occupy a hegemonic position in its own right and had to rely on additional support from rural elites to achieve their goal of leveraging sources of accumulation from the countryside. Chatterjee (1998) has described the post-Independence situation in India in terms of the Gramscian concept of 'passive revolution': industrial capital was advancing but had to rule in a kind of loose coalition with other dominant classes, making its advance a slow and uneven one.

India thus had a range of challenges to achieve its 'agrarian transition'—challenges which were only complicated by the size and diversity of the rural population.[4] Complicating matters further, there were conflicting views on the processes by which food production should be increased. Nehru favoured a policy of land redistribution. It was hoped that providing land to the tiller would give peasants an incentive to invest more of their labour in production and thus increase per acre output. Yet, this was not the view of other major factions within the Congress Party, many of whom were either from the landlord class or dependent on this class. As land redistribution threatened the landlord class with the loss of their substantial land holdings, these factions instead favoured policies that provided price incentives and technology to boost productivity.

Popular support for Nehru's leadership ensured that his redistribution framework won out in the short term and attempts were made to implement it throughout the nation. It was, however, a fraught process. As Herring (1983) outlines, land reforms in India proceeded in two phases. The first involved the abolition of landlordism, which directly attacked the interests of the parasitic *zamindar* class, who had acted as intermediaries between rural India and the British Raj. The 'Abolition Acts' were to appropriate the substantial holdings of absentee landlords, with the exception of the family farm, and redistribute them to the tiller. This was a costly process for the state and an incomplete one—many landlords found

[4] The first census of Independent India, conducted in 1951, found the rural population to be 82.7 per cent of total (Census of India, 1954). The rural population in India remains high by global standards, at 68.8 per cent as of 2011 (Census of India, 2011), highlighting the ongoing significance of some form rural consent to any national hegemonic regime.

legal loopholes to retain holdings. The second stage, which began in earnest in the 1950s, involved the implementation of 'land ceilings', in which landholdings above a certain size were to be redistributed to marginal farmers, the landless and agricultural cooperative societies. The Central Government justified this not only in terms of social justice but also in terms of the inefficient nature of larger farms. State governments, however—which were far more under the influence of elite rural classes than the Centre—were highly resistant to implementing the directives to impose land ceilings—and farmers with large landholdings repeatedly stalled implementation at the local level. Thus, while there was some progress in abolishing landlordism in the late 1940s and early 1950s, progress on enforcing land ceilings was extremely limited.

Though they did not achieve many of their core objectives, India's land reforms significantly altered the balance of social powers. Byres (1981) notes at least two outcomes of land reform on the class composition of rural India. First, large absentee landlords were almost completely liquidated as a class, having lost considerable legitimacy through the social justice rhetoric of the Independence movement.[5] Second, land reforms saw the consolidation of a class of capitalist farmers with small- to medium-sized holdings. Fearing the extension of land ceiling policies, a number of medium-sized absentee landlords became direct cultivators, with a direct interest in expanded production. Byres (1981) suggests that, as these emergent capitalist farmers consolidated their position economically, they also developed their capacity to act in their own political interest (developing as a 'class-for-itself'). In practice, this meant preventing further land redistribution. The influential rural elites within the Congress Party had already managed to dilute Nehru's policies and interfere with their implementation (Varshney, 1995), and larger landholders had used their connections at the local and state level to stall the process (Frankel, 1971). The emergent capitalist farmers only contributed further to this disruption of land redistribution. These factors prevented the success of the policy in its aims of increasing production, such that, after a modest increase during the 1950s, by the early 1960s agricultural output had stagnated and even dropped slightly (Varshney, 1995: 42–43). In policy circles it was felt that the subsequent inflation of food prices was stalling national economic growth (Frankel, 1971: 4–5).

In addition to land reform, Nehru's government also drew on another strategy to increase agricultural production, which it termed 'community development'. This was largely inspired by Gandhian ideas of transforming social relations within

---

[5] They did, however, use their political influence to secure generous compensation for land lost, which many used to begin careers in politics, ensuring continued political power for rural elites (Byres, 1981).

villages from below, particularly by encouraging local participation in development and the formation of village or district-level cooperative societies. As Sinha (2008) outlines, although this strategy was intended to provide empowerment for the rural poor, in practice, it tended to reinforce existing power relations. Project officials were reluctant to challenge village-level power structures, adhering to the notion that they should transform the *entire community* and not just the poorer sections. In practice, this meant allowing the perpetuation of the status quo. Given that many local-level community development workers tended to be from higher castes and worked more easily with their own caste members, in some instances, they served to amplify existing inequalities. Sinha also notes that community development provided an entry point for foreign institutions, such as UNICEF, the FAO, WHO, and the Ford and Rockefeller Foundations, which the Government of India drew on as advisors for the programme. These advisors ultimately became nested within the Indian state and came to exert considerable influence over subsequent development policy—particularly in guiding it towards more technocratic approaches to development and away from land reform.

In terms of hegemony, the land reform project failed to win the consent of influential rural elites and the emergent class of capitalist farmers, who used their power at the local level to stall and interrupt the implementation of reforms. With a few notable exceptions, small-holding peasants and landless labourers—who stood to benefit from the reforms—did not organise as a powerful bloc to ensure the project's success (Weiner, 1962). Indeed, in some cases, the structure of local social relations was such that the peasantry was materially and morally dependant on local landholding elites, discouraging them from actively pushing for land redistribution (Herring, 1981). Thus, as Varshney (1995: 46) puts it, the project of land reform lacked 'political microfoundations'. As a hegemonic project, land reform lacked the necessary dynamism, due to the dissent of elites and lack of active consent (participation) of potential beneficiaries. This provided an opening for Nehru's opponents to introduce technocratic policies as an alternative to land reform—policies which were supported by international institutions, which had become influential within the Indian state, partly using 'community development' as an entry point. The adoption of these policies came to be known as the Green Revolution.

Origins of the Green Revolution

The term Green Revolution refers to the rapid uptake of a package of agricultural technologies and practices, which occurred not only in India but in various parts of the developing world in the mid-twentieth century. At the heart of this package were the newly-developed high yielding variety seeds (HYVs), which had originally been produced by Western agricultural scientists working for large American philanthropic foundations. Though promising great increases in yield, the success of HYVs was highly dependent on inputs of synthetic nitrates and larger quantities of water than was used by traditional varieties. Consequently, they required access to modern irrigation systems to deliver their promised yield increases. Furthermore, the new varieties were generally more vulnerable to pest attack and disease, so synthetic pesticides were also promoted as a part of the Green Revolution technological package. This new approach would transform the way agriculture was conducted throughout much of the developing world, from the complex and labour-intensive land management strategies of traditional agriculture, to a capital-intensive approach involving mechanisation and the application of chemical inputs. The adoption of this new technological paradigm would also have a huge impact on agrarian social relations. In particular, it induced agrarian classes to become more dependent on the market.

The shift from land reform policies to Green Revolution in India is partly explained by the national hegemonic balance, discussed previously. A class of emergent capitalist farmers had become increasingly politically influential and had been lobbying government to create new commercial opportunities for several years (Baker, 1984). Yet, the picture is incomplete without also considering the global power dynamics involved. Patel (2013) persuasively argues that the Green Revolution was part of a longer-term global project to develop an integrated 'global food regime'. Led largely by the US government and capitalist philanthropic organisations, this food regime would be structured to allow for sustained capital accumulation for global agri-business firms, particularly through incorporation of Third World peasants into global food markets. In this sense, Patel suggests that at a global level, the Green Revolution must be seen as part of a much more long-term and ongoing project of accumulation, whose origins extend at least as far back as the colonial period.

The first large-scale deployment of Green Revolution technologies in the developing world occurred in Mexico in the late 1940s. The project was initiated by the Rockefeller Foundation, who, at the request of the Mexican government, sent a team to develop new varieties of crops. The team, led by Norman Borlaug, ultimately developed a dwarf variety of wheat, which was highly responsive to the application of synthetic nitrogenous fertilisers. While the increases in production the new crops facilitated were hailed as a major technological innovation, they were perhaps even more important to the reconfiguration of power and hegemony in Mexico's agrarian sector. Patel (2013) outlines some of the political and economic factors that were behind this intervention. The Rockefeller Foundation was highly connected to policy-makers within the US government and with global agri-business firms. There was a common concern amongst capitalist charitable foundations and the US government at the time with the rise of communism throughout the developing world, as well as peasant movements that were demanding land redistribution. In Mexico, there had been popular upheavals over the question of land reform, with ongoing conflict between the poor peasantry and the traditional elites. It was hoped that introducing commercial, technology-based agriculture could provide an alternative to ostensibly anti-capitalist land redistribution, and in a way that would be profitable to US-based agri-business firms. Indeed, the very term 'Green Revolution' was coined as a contrast to the 'Red Revolutions' that were underway in many parts of the world at the time.

The class dimensions of Mexico's Green Revolution must be emphasised. As would later be the case in India, Mexico's Green Revolution was introduced in the context of stalled land reforms, stagnant food production, and economic imbalances between agriculture and industry (de Alcantara, 1974). Importantly, the wheat variety was not distributed to the poor majority of farmers in the country (who mostly farmed beans and corn) but rather to wealthy land owners in the more extensively irrigated districts of the country (de Alcantara, 1974). The government formed extension agencies to promote the new technologies, whilst simultaneously banning the pre-existing peasant support groups, which had until then worked to assist poorer communally managed farms to reach markets. According to de Alcantara (1974), the result was that the communal farms gradually became less competitive and small-holding and sharecropping peasants became politically disempowered. Meanwhile, the emergent capitalist farms became increasingly profitable and politically powerful. For the most part, the new technologies did not reach the eighty per cent of Mexico's farmers who worked on small or communally managed properties, as they either could not afford them or they lived in regions where insufficient irrigation made them redundant. Thus, while Brown (1970) notes that by the end of the 1960s there had been a doubling and tripling of production of corn and wheat, respectively, de Alcantara (1974) points out that in the poorer rural areas, hunger and malnutrition persisted. The increase in food production (and subsequent decline in food prices) benefitted the urban population, while the rural peasantry, most of whom had been left behind by state development policy, suffered a decline in purchasing power, which prevented them from accessing this new food supply. Thus, while Mexico's Green Revolution was able to resolve issues relating to food prices in the cities, it could not resolve the inequalities and tensions between urban and rural societies. Similar patterns would later be observed in India and other developing countries who followed the Green Revolution path.

The sale of HYV seeds in Mexico proved immensely profitable for the Rockefeller Foundation and American agri-business, and by the 1960s, they sought to replicate their results elsewhere in the developing world, specifically Asia. Teaming up with the Ford Foundation, they jointly established the

International Rice Research Institute (IRRI) in the Philippines in 1960, for the development of HYVs of the staple crop. The choice of the Philippines was by no means arbitrary. As Patel (2013) outlines, as had been the case in Mexico, the US-based foundations partnered with the Philippines government in its HYV rice programme and used it as a basis to provide an alternative to escalating demands for land redistribution. Stemming calls for land redistribution in the Philippines was particularly important for the Foundations, who were concerned with the spread of agrarian communism in Asia (particularly China). They were also concerned with the prospect of the large estates in the Philippines being subdivided, as these had been cheap and reliable suppliers for US-based agri-business firms since the times of US colonialism (Putzel, 1992). The project thus followed the Mexican example, in simultaneously providing economic benefits and serving the *hegemonic* function of effectively defusing land redistribution as a pathway to development. In effect, it secured the interests of globally hegemonic groups (in this case, US agri-business) by simultaneously displacing all opposition and providing leadership for a model of agrarian development that focused on increased production, rather than challenging the distribution of resources. This pattern, Patel (2013) argues was characteristic of the Green Revolution project throughout the world.

IRRI proved to be tremendously successful, such that by the end of the 1960s, the use of the new seeds and the corresponding inputs had spread across the continent, with particularly rapid uptake in Pakistan, India, and the Philippines and also made substantial gains in Iran, Turkey, and Sri Lanka (Brown, 1968). Brown (1970) suggests that three factors were responsible for the rapid uptake. First, corporations, nation states (the US in particular), and philanthropic foundations were quick to invest in research in the field, after witnessing the immense profits made from the sale of seeds in Mexico. This allowed HYVs to be developed for Asia at a rate much faster than they had been for Mexico. Second, research groups, IRRI in particular, had multinational boards with connections to research networks throughout the world, allowing their findings to be quickly disseminated. Third, multinational corporations worked collaboratively with the research groups to ensure that the new inputs (fertilisers, machines, etc.) were available to farmers when the new seeds were released. In addition to these factors, politicians in post-colonial states saw the new technologies as a means of facilitating 'agrarian transition' in a way that bypassed the need for politically contentious land reform. By increasing food production and thereby reducing prices and creating taxable surplus income in the agrarian sector, states could effectively transfer wealth from the agrarian sector to fuel industrialisation. As such, there was a strong incentive for post-colonial states to promote the technology within their own countries. There was, thus, a convergence of interests and a kind of hegemonic pact between post-colonial states (and the emergent hegemonic pact of industrialists and agrarian capitalists within those states) and transnational actors—particularly American agri-business and 'big philanthropy'. The net effect of their collaboration was that by 1969 several Asian nations had returned to food self-sufficiency after a period of dependency, and investors in seed research and chemical inputs had made a fortune (Brown, 1970).

The Green Revolution in India

The implementation of the Green Revolution in India, which began in earnest in the mid to late 1960s, did not happen overnight. The foundations were laid as early as 1960, when the government, concerned by the slow pace of land reforms, introduced the Intensive Agricultural Districts Programme (IADP). Seven districts were selected in the initial stages of the IADP to receive extensive administrative support to adopt the new technologies. It was hoped that the technological progress in these districts would set the pace for technological development in agriculture throughout the country (Desai, 1969). In sharp contrast to the approach taken in Nehru's land reform project, in which backward districts were prioritised, participation in the IADP was awarded to districts with extensive irrigation facilities, thus benefitting regions that already had relatively high levels of agricultural development (Frankel, 1971: 3–5). The IADP was financed by the Ford Foundation (Desai, 1969), which, along with the Rockefeller Foundation, had also funded research centres throughout the country (Varshney, 1995: 77). Having learned from the Mexican experience, the Foundations knew they could not rely on the idea that the Green Revolution technologies were beneficial to small holders to mobilise a mass-base of beneficiaries, as the technology clearly tended to benefit larger, commercial landowners (Patel, 2013). Instead, they used their research centres to train large numbers of agronomists throughout India, convincing them of the merits of the technologies for boosting agricultural production. As such, when the time came to 'scale up' the model adopted in the IADP, much of the necessary moral and intellectual support was already in place.

In the mid-1960s, India was not as much under the influence of the US government as were Mexico and the Philippines. Domestic power dynamics are thus as important in understanding the manner in which its Green Revolution was implemented as international influences. When Nehru died in 1964, a power struggle ensued within the Congress Party between those who continued to support land reform and those who favoured a technocratic approach to agricultural development (Varshney, 1995: 48–70). The technocrats made the case that using redistribution alone would not allow production to increase at the speed necessary to meet requirements. Their case was strengthened by the gradual withdrawal of regular food aid from the United States and the impact of two successive years of drought in the mid-1960s (Cleaver, 1972; Varshney, 1995: 74–75). Furthermore, a technocratic approach would be more agreeable to middle to large landholders, who had interrupted the land-reform process, particularly those who were emerging as capitalist farmers and recognised potential profits in the new technology (Byres, 1981). Thus in 1965, the government initiated a new scheme, the Intensive Agricultural Areas Programme (IAAP), which would extend the approach taken in the IADP to one hundred and fourteen of India's three hundred and twenty-five districts (Frankel, 1971: 5). In addition to these measures to increase the use of the new technologies in agriculture, a policy was also implemented to provide minimum support prices on certain agricultural commodities, which gave farmers an incentive to invest in increasing their output (Varshney, 1995). Implementing the programme involved channelling credit and inputs to all regions with secure access to irrigation canals or tube wells. State governments and various organisations were encouraged to set up demonstration farms, which would directly demonstrate the ways in which the new approach could increase yields, and any special techniques needed to ensure success in local conditions (Sen, 1974).

The unfolding of the project was guided by hegemonic considerations. The emergent capitalist farmers, who had consolidated their political power in the land reform process, effectively guided the project towards districts in which they were in ascendency (Byres, 1981). With this, and in the development of support price policies, their perspectives began to be incorporated within the state and they began to articulate hegemony within the agrarian sphere. Furthermore, intellectual infrastructure was set in place to secure the consent of other agrarian classes for the new technology. A crucial component of this was the agricultural universities, one of which would be established in every state. The agricultural universities had two functions: first, to conduct research on HYVs and the chemical agronomic schedules best suited to local conditions for each state, and second, to conduct extension programmes to educate farmers on how to apply the new approaches on their farms (Randhawa, 1974). They served to establish consent for the Green Revolution and to provide it the prestigious aura of science. Indeed, as Gupta (1998) argues, the participation of peasants in the Green Revolution can in some measure be attributed to the circulation of development discourses that these institutions promoted, which positioned everything traditional as 'backward' and inferior and the Western approach embodied in the new technology as a sign of 'progress'.

In the districts targeted by the IAAP, the transformation was rapid. Many of the preconditions for the success of the Green Revolution had already been established under the British Raj, including the institutionalisation of private property rights and the penetration of state bureaucracies to local levels (Ludden, 1999). The two states that adopted HYVs most rapidly were Tamil Nadu and Punjab, which by 1970 had seventy-six and ninety-six per cent of their land holdings covered by

HYVs, respectively (Sen, 1974). This was due largely to the fact that they both had high-quality, reliable irrigation systems, and fertile alluvial soils. Punjab also benefited from having already undertaken a massive project in land consolidation, having good communication networks that allowed technology to diffuse rapidly, and a state government that was committed to boosting agricultural production by establishing various subsidies and other economic incentives for change (Randhawa, 1974; Singh and Kohli, 2005). Other regions with established irrigation systems, notably Haryana, Western Uttar Pradesh, and several isolated districts in South India, also saw radical changes.

In regions where there was no assured access to irrigation, however, change came very slowly, if at all. For Sen (1974), this was the major limitation of the Green Revolution approach in India. He argues that those who had anticipated far-reaching, national change had underestimated the extent to which HYVs depend on large, controlled doses of water, or overestimated the capacity to extend the area under modern irrigation systems. By the 1970s, it was apparent that the potential to expand the area under irrigation was extremely limited, due to the dry, semi-arid climate which pervades much of the subcontinent. To this day, there are many parts of the country that the Green Revolution never reached. Given agro-climatic conditions, Green Revolution technologies would not be appropriate in those regions. Nonetheless, regions in which the technologies were applied were radically transformed, not only in terms of agricultural output but also socially, culturally, ecologically, and economically. Regional disparities in agricultural production drastically increased, such that some regions became 'food bowls' for the nation, whereas others were left behind in terms of rural development.[6]

The Green Revolution was not only a package of technologies, it also relied very heavily on market supports, particularly subsidies on chemical inputs and minimum support prices for agricultural commodities. While farmers took a risk in purchasing the new technologies and agricultural inputs that the Green Revolution required, these supports provided some level of security. Indeed, Patel (2013) argues that, at least in the early stages, the Green Revolution *could not have functioned without these supports*. As farmers in Green Revolution zones became more integrated into the market, their interests coalesced around increasing the subsidies and minimum support prices for agricultural commodities. By the 1970s, farmers' movements began to form in these regions, composed largely of capitalist farmers with medium sized holdings and made further increases to these market supports their chief political demand (Gill and Singhal, 1984; Omvedt, 2005; Varshney, 1995).

[6] The state of Punjab, for example, though representing only 1.5 per cent of the geographical area of India, was producing one quarter of India's grain requirements by the early 1970s (Randhawa, 1974: 178).

The internal dynamics of these movements are worth noting, as they illustrate how a certain hegemonic configuration was maintained that allowed the class interests of capitalist farmers to prevail throughout the Green Revolution period. In a detailed study of the Bharatiya Kisan Union (Indian Farmers' Union, BKU) in Uttar Pradesh, Gupta (1998) demonstrates how farmers' movements were led by relatively privileged middle-caste capitalist farmers who used the movements to maintain a hegemonic position in the countryside. These leaders claimed to speak on behalf of the rural community as a whole, mobilising powerful populist discourses of a rural *Bharat* (the ancient name of India) opposed to urban *India*. They mobilised popular hostility towards urban classes including traders, money lenders, and revenue collectors who, they claimed, were plundering the countryside. Leaders would wear rustic clothing and present themselves as simple and economically weak, even when this was not the case, to signify their solidarity with the rural poor. More substantially, their politics revolved around lowering the costs of chemical inputs and raising minimum support prices and were thus squarely in the interests of capitalist farmers. By claiming to speak on behalf of all of rural society, however, they effectively marginalised the voices of rural subaltern classes and channelled rural dissent down acceptable channels. They were crucial to ensuring that social change in regions that followed a Green Revolution path of development would continue to benefit this rural elite. This kind of rural populism would continue to be influential in subsequent decades and is important in understanding how some discourses surrounding sustainable agriculture are framed (in particular, Chapter 4 of this book).

These movements should be taken as a key sign that the Green Revolution was, at least in some regions, operating hegemonically. The movements provided a mechanism for rural populations to express their dissatisfaction with some of the outcomes of the Green Revolution, whilst channelling dissent into demands that were harmonised with hegemonic interests (namely, the further development of intensive, capitalist agriculture). Farmers had some capacity to bargain for better conditions within the prevailing order, but little scope to raise questions about the fundamental assumptions of that order. In this way, by the 1970s, a hegemonic equilibrium[7] formed, within which even the appearance of conflict was actually a sign that hegemonic powers had won the consent of capitalist farmers (and those that followed them) for the Green Revolution paradigm.

[7] Note that for Gramsci (1971: 182) hegemony is achieved when 'the dominant group is coordinated concretely with the general interests of the subordiante groups, and the life of the State is conceived of as a continuous process of formation and superseding of unstable equilibria ... between the interests of the fundamental group and those of the subordinate groups—equilibria in which the interests of the dominant group prevail, but only up to a certain point'.

To summarise, for the Indian state, the Green Revolution served at least three functions. It was (1) a source of capital accumulation, as the new technology made both chemical input industries and commercial agriculture more profitable (at least in the short term); (2) a way to increase food production; and (3) it resolved an impasse for the Indian state, in which the need to increase food production through land reforms had been stalled by the lack of consent of medium to large capitalist farmers. In this impasse, the Green Revolution performed a critical hegemonic function: it allowed the Central Government to win the consent of emergent capitalist farmers for their approach to rural development. This was critical, since this class had considerable and growing influence within the ruling Congress Party. Indeed, the emergent class of capitalist farmers were effectively incorporated within the hegemonic bloc within the Green Revolution period, and by the late 1970s had considerable control over the state (Byres, 1981; Kaviraj, 1986).[8] This class of farmers, and the new intellectual infrastructure of extension services and agricultural universities, articulated hegemony at the local level, by enlisting the consent of smaller landholders for a form of agrarian development that bypassed land reform and yet still provided some benefits to farmers with various sized holdings. Crucially, the project also harmonised with the interests of powerful transnational actors, in particular global agri-business, 'big philanthropy' and the US government, who saw it as an opportunity to defuse 'communist' demands for land redistribution in Asia and establish a food regime which would be immensely profitable to US-based transnational agri-business. As such, they provided various forms of material and ideological support. As Frankel (1971) details, nowhere in India was the implementation of the Green Revolution a smooth process, as there were various forms of resistance from agricultural labourers and marginal farmers. As time passed, however, the process eventually stabilised, with consent crystallising around a commercially-oriented approach that, in large parts of the country, involved high levels of inputs (supported by subsidies) and outputs.

The aftereffects and decline of the Green Revolution

Since the early stages of its implementation, there were concerns regarding the impacts of the Green Revolution. First, there were concerns regarding the uneven distribution of benefits of the new technology. As Frankel (1971) outlines, the profits made by small holders from the new technology tended to be consumed by the increased cost of inputs and living expenses, while for large land holders even

[8] To illustrate, Prime Minister Chaudhary Charan Singh, who served in office from 1979–1980, had been leader of a powerful farmers' movement in Uttar Pradesh, whose legitimacy derived from his claims to represent the entire rural community, though in practice tending to represent the interests of wealthy capitalist farmers (Byres, 1981; Gupta, 1998).

small increases in output on a per acre basis added up to substantial increases in aggregate income. Larger landholders also had more surplus capital when the new technologies were introduced, allowing them to invest in tube wells, fertiliser, and mechanisation, which ensured that the benefits of the new varieties of seeds were realised. Small holders were often unable to invest in the full package of Green Revolution technologies and thus made only modest gains. A significant number of small holders became uncompetitive and were forced to sell their land. Large land holders were eager to acquire this land to reap higher rates of profit, and their demand drove up land prices by as much as 500 per cent in parts of the country, making land more inaccessible for the poor majority (Cleaver, 1972).

Disparities in benefits were even more pronounced in the case of agricultural labourers. In the initial stages, regions that followed the Green Revolution development trajectory did see a rise in on-farm employment and a subsequent improvement in the bargaining power of agricultural labourers (Bhalla, 1976; Muttiah, 1970). Nonetheless, Frankel (1971: 105–109) suggests that despite these short-term benefits, wages did not undergo sustainable increases, as may have been expected. Because of the growing regional disparities introduced by the Green Revolution, labourers migrated from less productive areas, creating surplus labour in the areas of high production. The outcome of this was that wages remained static, while the cost of living increased. Furthermore, while on-farm jobs increased in the short term in intensively cultivated regions, as mechanisation gathered pace, jobs became fewer, more seasonal, and short term in nature (Byres, 1972). Patel (2013: 25–26) argues that these conditions were exacerbated by the state in Green Revolution regions, which provided disproportionate support to land holders and discouraged the organisation of labour. As a result of these factors, agricultural labourers received a very meagre portion of the share of wealth generated by the Green Revolution (Bardhan, 1970).

These growing social and economic inequalities had political consequences. Just as Bates (2005) had observed in Africa, certain agrarian classes in India had become beneficiaries of the new model of agrarian development, removing the potential for any unified opposition from the agrarian sector to the state's overall development strategy. The Green Revolution had created a wealthy class of capitalist farmers, who were the chief beneficiaries of the process. As this class attained political power from the village to the national level, they ensured that the benefits of future policies would also flow on to them (Byres, 1972). The political leadership of these classes helped to develop consent for the Green Revolution.[9] As

[9] As is always the case, however, consent was not complete. For example, in the late 1960s, communists in East Thanjavur District in Tamil Nadu mobilised landless labourers against the decline in wages and increased cost of living that had accompanied the Green Revolution,

Varshney (1995: 81) puts it, where previously agrarian mobilisation occurred on the basis of class, this new leadership attempted to mobilise the entire agrarian sector, through demands for more favourable subsidies and minimum support prices. At least in the early 1970s, there was no effective leadership to unite those disaffected by the Green Revolution (Byres, 1972; Cleaver, 1972). Yet for Byres (1972), this hegemonic equilibrium could not endure in the long term, as the concentration of political power in the hands of wealthier farmers prevented the contradictions of capitalist agriculture from being resolved.

By the late 1970s and early 1980s, it became apparent that the economic gains that had occurred in the first ten years of the Green Revolution could not be sustained. Long-term studies have shown that yields for wheat and rice in Green Revolution zones have stagnated and, in some cases, declined. Numerous explanations are provided for this, with some studies suggesting that the most enterprising farmers found it more profitable to invest surpluses outside of agriculture (Baker, 1984) and others suggesting declines in yield are associated with a loss of soil nutrients (Bandhari *et al*, 2002; Ladha *et al*, 2003). Whatever the reason, it appears that yield increases through Green Revolution methods have reached their limit. This has had a range of social and economic implications. Since the price of inputs continues to increase[10] and food prices have remained relatively low,[11] the profitability of agriculture as an economic venture has declined over time (Sidhu, 2002, 2005). Gill (2005) notes that even with government

which ultimately resulted in deadly confrontations with landlords and police (Alexander, 1975a; Mencher, 1974). While such confrontations did occur in various parts of India in the early phases of the Green Revolution, they were mostly localised, lacking national or state level organisation or any alternative hegemonic project. West Bengal provides the clearest exception to this, where, in response to strong, consistent mobilisations from the rural poor, the Communist Party of India (Marxist) reintroduced land reform in the 1970s (Kohli, 1987).

[10] Synthetic fertilisers depend upon fossil fuels for their production, and so the price of fertilisers is highly sensitive to changes in the oil price. The actual cost of fertilisers has risen sharply in the last decade, and with the spectre of peak oil, this upward trend is likely to continue. For the time being, farmers in India have not directly borne all of this increase, due to subsidies on inputs, but the government expenditure on subsidies is beginning to reach unsustainable levels (Gopikrishnan, 2012).

[11] Although global food prices have undergone a series of surges since 2006, this has generally not led to an improvement of the economic position of farmers. As Headley and Fan (2008) argue, much of the food price increase is attributable to the rising oil price, which raises the cost of production. Further, they suggest that a large portion of rural households are net food consumers, with diversified income, and are thus as vulnerable to food price increases as the rural poor.

subsidies for fertilisers and pesticides, input costs are consuming an ever-greater proportion of farmers' disposable income, even in relatively privileged regions like Punjab and Haryana. For small holders, this has made farming unviable, and many have been forced to sell their land. This has implications for employment, as larger farmers tend to employ fewer workers per acre. For large land holders, declining profitability means a loss of capital for investment in rural areas, reducing prospects for the creation of employment, both on and off farms (Chand, 2009). The development of these issues in Punjab, which was once considered the exemplar of Green Revolution 'success', has led Sidhu (2002) to argue that the rural development model pursued in the region should now be considered a prime example of economic and social 'non-sustainability'.

Over the long term, the Green Revolution has also resulted in serious ecological damage, raising further questions as to its sustainability. In the late 1970s, Yapa (1979) raised concerns that the neglect of ecological issues within the Green Revolution paradigm would have serious consequences for the future. He suggested that the vulnerability of HYVs to pests would lead to the use of an ever new range of pesticides, the strength of which would have to be incrementally increased as pests became resistant. Studies from India have confirmed these predictions, with high levels of resistance to numerous pesticides being registered in various parts of the country from as early as the 1960s but substantially increasing by the 1980s (Mehrotra, 1989). This, in turn, led to an increase in pesticide usage, which not only was an additional financial burden on farmers but also caused further ecological contamination and damage to non-target animals, including humans. The effect of this has been exceptionally pronounced in regions where highly intensive chemical agriculture has been practiced. In Punjab, there is a high level of pesticide contamination in water, and unsafe levels of pesticide residues have been found in a variety of foods produced in the state (Tiwana *et al*, 2009). Epidemiological studies have now linked the usage of pesticides to the high rates of cancer (Kochhar *et al*, 2007; Thakur *et al*, 2008) and reproductive health disorders (Thakur *et al*, 2010) in parts of the state.

Several other environmental concerns have been raised. Yapa (1979), for example, highlighted the potential of damage to soil structure from excessive use of synthetic nitrates and loss of biodiversity through mono-cropping of HYVs. In Haryana, which has had a similarly intensive cropping pattern to Punjab, a range of environmental issues have been observed, including chemical contamination of water systems, salinity, soil erosion, desertification, and water logging (Singh, 2000). The damage to soil can be attributed to the loss of tree cover and the fact that the intensive monoculture of the wheat-rice cropping pattern has caused nutrient deficiencies and exhausted much of the ground water supplies (Singh, 2000). Indeed, the loss of ground water is becoming an acute issue in Punjab, Haryana, and Rajasthan, with NASA satellite scans revealing mean declines in the water table of up to four centimetres per year (Rodell, Velicogna, and Famiglietti, 2009).

These long-term degenerative patterns have undermined the legitimacy of the Green Revolution in India. When seen in combination with the impacts of liberalisation on India's rural economy, the situation may be viewed as an 'agrarian crisis', and even a 'crisis of authority' in which subaltern groups are no longer able to lend their active consent to the hegemonic bloc.

Liberalisation and the Agrarian Crisis

The deterioration of conditions in the countryside has been aggravated by the liberalisation of India's economy. Before liberalisation, India had followed a planned approach to development, with import substitution policies aimed at developing and protecting domestic industries. Although modest moves to expose India to the global economy had occurred in the 1980s under Prime Minister Rajiv Gandhi (Corbridge *et al*, 2013), liberalisation began in earnest in 1991 when, following a balance of payments crisis, India accepted a US$ 4.8 billion loan from the International Monetary Fund. The loan was received with the condition that India make widespread policy reforms, including the removal of barriers to trade, deregulation of the finance sector, and the privatisation of public assets (Patnaik and Chandrasekhar, 1995). Among the most immediate impacts of liberalisation was an increase in rural poverty (Patnaik and Chandrasekhar, 1995: 3007). Lerche (2011) argues that liberalisation made the Indian economy less dependent on the agricultural sector. Not only did it enable the import of cheap food from overseas but the availability of transnational finance reduced the importance of rural India as a source of capital. For this reason, Bernstein (2009) suggests that liberalisation has allowed post-colonial states to dodge Byres' (1991) 'agrarian transition' (the transfer of resources from the agrarian to the industrial sector) and the complex hegemonic manoeuvring it entails. Being less dependent on the agrarian sector, support for agriculture has become a lower priority for the state,[12] and, as is consistent with neoliberal ideology, subsidies and support prices have been lowered or removed.

The most direct consequence of these developments was a decline in agricultural growth (Reddy and Mishra, 2009b).[13] Patnaik (2003) argues that it would be a

---

[12] This decline in the priority status of agriculture for the Indian government is reflected in expenditure on agricultural development, which, after liberalisation, fell from 13.2 per cent of GDP to 7.8 per cent (Patnaik, 2003: 48).

[13] '[T]he growth rate of agricultural GDP decelerated from 3.08 per cent during 1980–81 to 1990–91 to 2.57 per cent during 1992–93 to 2005–06 ... [T]he growth rate in foodgrains ... fell from 2.85 per cent in the 1980s to 1.16 per cent in the 1990s, lower than the rate of

mistake to assume that this slowing of growth and corresponding rise in rural poverty was simply an unintended by-product of liberalisation. Rather, rural poverty was the immediate and predictable consequence of policies intended to cause deflation, to create a more profitable environment for finance capital. In the period from 1995 to 2001, all quantitative restrictions on food imports were removed, which, combined with other deflationary policies, caused a drop in the price of all agricultural commodities (Patnaik, 2003: 40–41). Consequently, the profitability of agriculture fell, as did livelihood security. The growth rate in food grains production fell by half during the 1990s, such that it was below the population growth rate for the first time since the introduction of the Green Revolution (Patnaik, 2003: 48). Cropping patterns throughout India also shifted, away from the immediate food needs of the local population and towards more profitable, export-oriented crops, such as cotton, which had an adverse effect on local food security (Patnaik, 1996).

The souring of the Green Revolution and the deleterious effects of liberalisation have now developed to such an extent that a number of scholars refer to the situation in rural India as one of 'agrarian crisis'. Jodhka (2012) suggests that after a decline in the study of agrarian change in India in the 1990s, from the early 2000s onwards, 'crisis' emerged as the principal discursive frame through which rural India was interpreted. Jodhka is critical of this trend, as it tends to be applied indiscriminately across the entire agrarian sector and is insensitive to class and caste differences within the sector. In some respects, this critique is misplaced, as some of the most prominent studies that have used the frame of agrarian crisis have examined the differential impacts of the decline in the agrarian sector on a variety of classes (Lerche, 2011; Patnaik, 2003; Reddy and Mishra, 2009a; Walker, 2008). Nonetheless, it is true that the term is often used without a great deal of precision, such that it is not entirely clear at what point rural India entered into a state of 'crisis' and what the implications of this crisis are.

From a macro perspective, India's agrarian crisis has been interpreted as a growing imbalance between rural and urban economies. As Reddy and Mishra (2009a, 2009b) outline, while the gap in Gross Domestic Product (GDP) between rural and urban India has increased, there has not been a corresponding shift in employment patterns. In essence, the growth in urban and non-agricultural rural employment opportunities has not been fast enough to accommodate the increasing number of people from rural communities who can no longer sustain themselves on agriculture alone. Since the number of people for whom agriculture remains the

growth of population of 1.9 per cent during the latter period' (Reddy and Mishra, 2009b: 11).

only livelihood option continues to grow, there has been an increase in the number of land holdings. Given the reduction in the absolute size of cultivated land in India (as a result of urbanisation, land degradation and so on) the increase in land holdings has meant a reduction in the average size of land holdings. Farmers must produce in more competitive, less profitable conditions with less land, leading to increasing precariousness.

The most alarming sign that these developments have brought agrarian India to a crisis point is the dramatic increase in farmers' suicides since the late 1990s. Studies of farmers' suicides provide an insight into cases of extreme desperation experienced throughout rural India: as Rao (2009: 111) puts it, they are the tip of the iceberg of rural distress. While the data on suicides is somewhat patchy, and many are unreported, official figures show an increase in farmer suicides from an average of 15,747 per year in the period 1997–2001 to an average of 17,513 per year for 2002–2006 (Sainath, 2009). Farmer suicides are clustered in certain regions, and studies indicate common features. In Andhra Pradesh, where a wave of farmers' suicides caught national attention in the early 2000s, Galab, Revathi, and Reddy (2009) found that most of the cases were farmers who had moved from subsistence agriculture to cash cropping, in the hope of improving their economic position. Cash cropping exposed farmers to considerable risk and many of them sustained heavy losses. Another common feature among farmers who committed suicide is indebtedness. The declining profitability of agriculture made farmers more reliant on credit and less able to repay loans. This came at a time when the rural credit market was being deregulated, reducing poor cultivators' access to institutional sources of credit, and often forcing them into highly exploitative relationships with local money lenders (Shetty, 2009). Mishra's (2009) research on farmers' suicides in Vidarbha, a region in Western Maharashtra with some of the highest suicide rates in the country, found that farmers who had committed suicide had debts 3.5 times higher than control groups. Studies from other parts of the country have had similar findings (Gill, 2005; Jeromi, 2007). Rural suicides, therefore, are not merely signs of individual psychological disturbances but are clearly linked to the effects of liberalisation and deteriorating conditions in the agrarian sector.

The 'agrarian crisis' referred to in previous studies reflects a scenario of pronounced distress in the agrarian sector with limited or no prospects of improvement in the foreseeable future. Indeed, when one considers both the ecological and economic dynamics of the situation, it appears that within the current framework the future trajectory can only be downwards. It is a 'crisis', and not merely a 'downturn', because the social and economic constraints, particularly on the lower class strata, currently appear intractable.

Agrarian crisis as 'crisis of authority'

Rural India clearly faces a 'crisis' in terms of growth rates, employment patterns, marginalisation, and ecological decline; yet, has this has led to a 'crisis of authority'? Have these material conditions led to a decline in the legitimacy of hegemonic groups and ideas? The Green Revolution was a source of capital accumulation, food stocks, and political consent. To a greater or lesser extent, all three of these functions have come into a state of crisis throughout much of rural India. The current situation can certainly be seen in terms of a crisis for capital, in the sense that the conditions for continuous accumulation in the agrarian sector have not been sustained. In the long term, this may also become a food crisis, as food production fails to keep pace with population growth and growing inequality creates issues of access to adequate, nutritious food. The idea of a crisis in political consent, however, is of most interest to the present study. In the wake of the Green Revolution, consent consolidated around a commercial model of agrarian development, held up by support prices and subsidies. Dissent was largely channelled through farmers' movements, led mostly by wealthier farmers, whose demands were for increases to these support mechanisms, rather than any fundamental change in the system. The developments described above indicate that the sustainability of the current approach to rural development is losing plausibility. The active consent of the rural population for the status quo is weakening, though no clear alternatives have emerged, raising the spectre of a 'crisis of authority' in which 'the old is dying and the new cannot be born' (Gramsci, 1971: 210).

Active consent has become difficult to maintain, as the means of social reproduction are under threat. The rates of growth brought about by the Green Revolution had provided a temporary solution to what Bernstein (1996) terms 'the agrarian question of labour'—that is, how agrarian classes can survive, given capitalism's uneven growth trajectories between agriculture and industry. Now that agricultural growth rates have sharply declined, this 'agrarian question of labour' re-emerges. The Green Revolution has lost its tenacity as a unifying discourse, and subaltern groups have a new impetus to articulate alternatives. As the Green Revolution approach (complete with its regional disparities) is still practiced, it would appear that this approach is merely 'dominant' rather than 'leading'—it remains in place purely for lack of perceived alternatives.

Liberalisation has been particularly disruptive to the complex hegemonic balance that formed under the Green Revolution. It has undermined some of the most crucial factors for maintaining consent for agrarian development since the 1960s—namely, stable prices for agricultural commodities, which allowed farmers to invest with good prospects for profitable returns, and market supports. The politics of farmers' movements had centred on various tactics of improving the terms of trade for agriculture. Farmers' movements lost much of their bargaining power upon entering a neoliberal political environment, which accepted the volatile prices of the global market. In response to this, a number of farmers' movements emerged with an explicitly anti-commercial, anti-liberalisation agenda, with the objective of fostering conditions for greater farmer self-reliance (Mukherjee, 1998; Omvedt, 2005). This included, most notably, the Karnataka Rajya Raita Sangha, which developed in the 1990s, and began to articulate a model of development that would be more sustainable for small-holding farmers (McHattie, 2000). The demands of these new farmers' movements are less amenable to neoliberal hegemony than those of the older breed. Yet theirs is not the only voice in rural civil society—a number of farmers' movements, such as those led by Sharad Jyoshi, take a pro-liberalisation stance, believing that Indian farmers are sufficiently versatile and entrepreneurial to profit from the new market environment (Omvedt, 2005). Farmers' movements have become highly divided, no longer uniting the agrarian sector as Varshney (1995) claims they had in the 1970s. Thus, while discontent exists and is widespread, there is no single group in civil society that is capable of uniting and representing this discontent, let alone consensus regarding an alternative.

Liberalisation has reconfigured hegemony in rural India. Saull (2012) suggests that liberalisation initiated a new historical bloc, with greater integration of all sectors into the global economy. Patel (2013) argues that a key component of this has been a far greater involvement by the private sector, with small agri-business firms standing to gain more than farmers. These firms play a key role in developing consent for liberalisation at the local level. Yet, it appears that the global hegemony of neoliberalism and finance capital is less capable of establishing genuine grassroots consent and tends to operate in a cruder, more dominative manner in the Indian countryside. The neoliberal state has demonstrated a greater tendency to resort to force when hegemonic interests are compromised.[14] This shift towards oppression implies a shift towards a 'crisis of authority'. There have been clear signs of such developments throughout rural India in the decades since liberalisation began.

Liberalisation has generated widespread resistance in India. Rather than attempting to accommodate this resistance (and hence maintain a more hegemonic presence in the countryside), the state's response to resistance has been largely

[14] Howson (2006) distinguishes different usages of the term 'hegemony' in Gramsci's work, contrasting 'aspirational' and 'dominative' types. While 'aspirational hegemony' is open to negotiation and the building of consensus with subaltern groups, 'dominative' hegemony relies more on coercion, particularly when certain non-negotiable 'hegemonic principles' are called into question.

coercive, involving combinations of violence, deception, and fraud (Sinha, 2015). The neoliberal state has prioritised creating profitable conditions for investors. As Walker (2008) argues, this has come into direct conflict with rural communities when the state has forcefully acquired land from farmers and transferred it, at little or no cost, to domestic and foreign capital. While the Indian state has always had the right (under the Land Acquisition Act of 1894) to acquire land for use that is in the 'public interest', since liberalisation the meaning of the term 'public interest' has been modified, so as to include direct transfer to a private corporation. As such, the rate of dispossession has dramatically increased (Walker, 2008). The coercive nature of land acquisition has become particularly apparent in the case of 'Special Economic Zones' (SEZs). In these geographically delineated sites of between 10–5,000 hectares, there is no democratic governance, allowing for experimentation with levels of liberalisation that would not be possible within an ordinary democratic setting (Levien, 2011: 454). Therefore, by their very nature, within SEZs, all democratic means of consent building have been suspended, with the implicit acknowledgement that such consent would be highly unlikely or would disrupt investment plans. As Levien (2011) outlines, since the legal framework for SEZs was finalised in 2005, hundreds have been approved throughout the country. The land for SEZs is compulsorily acquired by the state from farmers with little or no compensation and transferred to investors at 'incentive' prices. There do not even appear to be gestural indications that the state is interested in building the consent of the rural community for these interventions.

Unsurprisingly, there has been widespread opposition from farmers to this coercion (Walker, 2008). In parts of Eastern India, in which agricultural underdevelopment and rural poverty were already extremely high, land acquisition has provided fuel for the ongoing Maoist insurgency. Maoists have forcefully ejected the state from districts in West Bengal, Chhattisgarh, and Jharkhand, establishing alternative systems of government and rendering large parts of the country ungovernable (Chakravarty, 2008). The Maoist insurgency can be seen as symptomatic of a crisis of authority, reflecting the fact that the state is unable to keep up with ground realities and re-establish some form of consent for the dominant system, despite the fact that such consent is clearly in its own interest. As Levien (2011: 455) notes, a number of commentators have suggested that the opposition of farmers could stall India's growth trajectory. Effectively, neoliberal India's neglect of consent from rural populations could undermine its own foundations as both an economic and hegemonic project.

To an extent, the state appears to have recognised the magnitude of the crisis and has implemented some responses to restore economic and political balance. One of the more progressive responses has been the National Rural Employment Guarantee Scheme (NREGS), which was introduced by the United Progressive

Alliance (UPA) Central government in 2005, in recognition that liberalisation had not generated livelihood options for the rural poor. The Scheme guarantees rural labourers 100 hours of work per household, per year, at the official minimum wage. Much of the work allocated has been directed to sustainable rural development, such as managing soil and water resources, and civil society organisations have played a strong role in guiding this allocation (Reddy and Upendranath, 2010). In this way, the Act not only has the advantage of addressing rural livelihood and development issues but it has a dynamic component, responding to issues and needs identified within civil society.[15] Despite these examples of the state attempting to establish consent, many of the other ostensibly 'progressive' policies seem to operate more as knee-jerk responses to a pervasive crisis. For example, Steur (2015) demonstrates how this operates in the case of India's 'Corporate Social Responsibility' (CSR) laws, which require corporations to 'give back' a proportion of profits to communities. She shows that in practice, CSR contributions are more often directed at fracturing opposition to displacement and dispossessed by selectively appeasing sections of communities, rather than establishing genuine consent for neoliberal development.

A more capital-friendly response to the crisis has come in the push for the introduction of genetically-modified (GM) crops. In 2009–2010, the UPA government sparked controversy by opening debate on the commercial release of 'Bt-brinjal'—a variety of eggplant, genetically modified such that every cell of the plant produces its own pesticide. Monsanto and Mahyco, the companies that produced the organism, argued that it would increase production by making plants more resilient against pests and reduce farmer dependence on pesticides (Barvale, 2009). Those opposed to the release of Bt-brinjal, however, have cited a number of risks associated with the technology, including the increased cost of seeds and its uncertain health and environmental impacts (Kuruganti, 2010). Responding to the controversy, the Central government's Environment Minister Jairam Ramesh hosted a series of public consultations, to gauge public opinion on the issue. The consultations showed widespread public concern regarding the new technology, and Ramesh responded by implementing a two-year moratorium on its commercial release (*The Hindu*, 2010). In implementing the moratorium, however, Ramesh was fundamentally at odds with other forces within the government, who remain committed to technological solutions to the agrarian crisis. Since assuming office

[15] It should also be noted, however, that the NREGS has had its detractors. Wealthy capitalist farmers and absentee landlords have objected to the impact of the policy in reducing the availability of labour and increasing the cost of wages, and have mobilised to demand the scheme be scrapped (Vakulabharanam, *et al*, 2011). Conflict over the NREGS reflects the challenges that the state faces in re-establishing consent in the current environment.

in 2014, the current Modi government has quietly attempted to resume field trials of a variety of GM crops, claiming that these are an essential component of India's 'Second Green Revolution' (*Reuters*, 2015).

As long as the state remains committed to neoliberal principles, it is difficult to implement any solution to the agrarian crisis that can address fundamental issues and re-establish the active consent of the rural community. Coercion remains the norm when it comes to managing rural discontent. While the NREGS has some merits, the GM 'solution' simply appeals to some of the most questionable assumptions of the Green Revolution discourse—namely that technological innovation will always allow food production to keep pace with population growth. There is an urgent need for a more complex, systemic view of the problem and innovative solutions that actively incorporate the needs and aspirations of the rural population. Such an approach is necessary not only to address the substantial social issues in the Indian countryside but also if there is to be any enduring political stability.

Conclusion

By taking a Gramscian perspective on India's recent agrarian history, this chapter has shown how the path of agrarian development that India has followed since Independence has closely reflected the hegemonic balance of power. The initial impulse to expand agricultural output came from hegemonic interests in transferring resources from agriculture to fuel the industrialisation process. The project of land reform was incomplete, as it failed to secure the consent of powerful rural elites and ascendant capitalist farmers. The Green Revolution, though causing significant instability in the short term, managed to broker a complex and relatively stable hegemonic balance in the 1970s, in which farmers' movements, led by wealthy capitalist farmers, negotiated increases in subsidies and minimum support prices, to sustain the security of their livelihoods. This ensured that some benefits of development remained in the countryside. The relative stability of this arrangement has since been disrupted, however, as the benefits of the Green Revolution expired and liberalisation has precipitated a widespread agrarian crisis. This crisis has undermined some of the key instruments for fostering consent among the rural population, raising the spectre of a Gramscian 'crisis of authority'.

The issues outlined in this chapter are significant to the present study for two reasons. First, they highlight the extent of the distress in rural India and the urgent need for alternative approaches to rural development. Second, they show that insofar as there has been a breakdown of political consent for the hegemonic food regime in India—thus constituting a crisis of authority—this breakdown also opens up 'space' in civil society within which alternatives can be propagated.

If discourses of Green Revolution and the old farmers' movements no longer hold influence over the popular imagination, then people may be more open to alternatives. This applies particularly to people whose experiences have made them disenchanted by the Green Revolution and liberalisation. Social groups who had previously been subaltern may attempt to take up leadership positions in this context, developing consent for alternative approaches. Alternatively, external agents may begin to work with these disaffected social groups to articulate a new hegemonic framework. The sustainable agriculture organisations that form the focus of this book are, in many ways, tangled up in these processes of articulation.

3

Embedded in Power

Potentials and Constraints of Sustainable Agriculture

In Chapter 2, I argued that the current situation in rural India may be described as a Gramscian 'crisis of authority'. Consent for dominant ideologies is breaking down, which provides a political opportunity for subaltern groups to assert alternatives. Yet, as Gramsci is keen to emphasise, in the analysis of such a crisis, the balance of power between different groups is of crucial importance. Not all groups are equally positioned to develop consensus for a new hegemonic formation. Indeed, Gramsci (1971: 210) warns of the risk that:

> The traditional ruling class, which has numerous trained cadres, changes men and programmes and, with greater speed than is achieved by the subordinate classes, reabsorbs the control that was slipping from its grasp.

The terrain of civil society, in which consent for new hegemonies is potentially established, is infused with power and privilege. If there is a 'crisis of authority' in the hegemonic approaches to agrarian development, powerful groups may be in a better position to respond to it. In a country such as India, the existence of multiple forms of inequality has a strong bearing on which groups are able to articulate alternatives to the current system and whose voices are heard. The voices of middle-class NGO workers with links to the development establishment often drown out those of others in civil society, which may ensure that hegemonic interests are not seriously questioned. Furthermore, at the local level, rural elites guide ongoing processes of change to ensure that they do not compromise their interests. Any attempt to formulate a sustainable and equitable response to India's agrarian crisis thus involves the navigation of a highly complex, uneven, and challenging terrain.

This chapter develops a conceptual framework to locate sustainable agriculture organisations within India's civil society, particularly attending to their position relative to various hegemonic forces. I begin by outlining the counter-hegemonic potentials latent within sustainable agriculture in the context of the present agrarian crisis. I then go on to outline the resource needs of sustainable agriculture organisations, highlighting that these needs often require external support. In contemporary India, this support is typically drawn from urban middle-class development workers. Not only have the urban middle class been historically significant in 'articulating hegemony' throughout India (Deshpande, 2003), but in the process of organising sustainable rural development projects, they become involved with various groups with vested interests. I pay particular attention to their relations with donors, the state, activist networks, rural elites, and the rural poor and outline how each of these agents may influence the outcomes of sustainable agriculture projects.[16]

Sustainable Agriculture as a Response to Crisis and Challenge to Hegemony

To a certain extent, a 'crisis of authority', a situation in which 'the old is dying and the new cannot be born' (Gramsci, 1971: 210), always involves the question of sustainability. The old hegemonic configuration cannot endure, and it is simply a question of what the new configuration of power will be. Yet 'sustainability' is likely to be the central organising principle of a new hegemonic formation in cases in which the unsustainability of the existing approach is clearly manifested on multiple levels. It is a situation in which the horizon of the future appears disastrous, and only an approach that remedies current trends towards decline can be perceived as acceptable (Brown, 2016). Arriving at a situation in which this is the predominant interpretation of a crisis *and* there is consensus on the kind of remedy that would be appropriate or acceptable is contingent on a good deal of persuasion and consent-building in civil society. Whether and how this occurs hinges on the existing balance of power between different groups. This highlights that sustainable agriculture is a politicised and contested concept—with different groups attempting to define it in ways that reflect their interests (Buttel, 1993). What is sustainable to one group may be deeply threatening to another, and developing a model that is suitable for an entire community will inevitably involve trade-offs (Conway and Barbier, 1990).

Many of the key factors in India's agrarian crisis concern issues of sustainability. When the economic trajectory is towards higher input costs and deflated agricultural commodity prices, the capacity for small holders to sustain themselves into the future becomes questionable. Policy-makers' preferred option is to encourage small holders to leave agriculture for urban employment, paving the way

---

[16] Sections of this chapter previously appeared in Brown, T. 2016. 'Civil society organizations for sustainable agriculture: negotiating power relations for pro-poor development in India.' *Agroecology and Sustainable Food Systems* 40 (4): 381–404.

for larger sized land holdings that can be farmed using industrial methods. Yet, in a country such as India, in which most of the population live in the countryside and there are few prospects for growth in non-agricultural employment (Reddy and Mishra, 2009b), the potential for doing this in a socially and ecologically sustainable manner that wins the consent of broad sections of the population seems unlikely. Sustainable farming models appear to offer a more immediate and accessible solution to agrarian crisis that appeals to the needs of resource-poor farmers. It reduces the inputs required to generate similar levels of output, while also addressing the ecological unsustainability of Green Revolution agriculture.

Sustainable farming practices integrate naturally occurring processes, such as nitrogen fixation and nutrient cycling, into farming systems. They make more efficient use of locally available resources and manage these resources more effectively (Altieri, 2002). Studies have found that such techniques, when appropriately implemented, can improve yields by up to 93 per cent (Pretty, Morison, and Hine, 2003). This can improve food security while reducing poverty. By reducing chemical inputs, it also lowers farmers' expenditure and dependency on agri-business for production. Studies of farmers' decisions to convert to ecological farming techniques suggest that, in addition to environmental and health concerns, farmers are often motivated by concern over the cost of chemical inputs (Fairweather, 1999; Darnhofer *et al*, 2003). With increasing price volatility and constrained access to cheap rural credit, farmers may perceive benefits in reducing their dependency on external inputs. In the future, peak oil's effects on input prices and projected shifts from subsidised chemical inputs to direct cash transfers to farmers are likely to make these benefits more apparent (Kapur, 2011).

Indeed, it is said that sustainable farming techniques can be a cornerstone for farmers in regaining 'food sovereignty'. The corporate food regime has resulted in heavy indebtedness, the pressure to integrate into global markets and overt state coercion to force people from land. Collectively, these have severely curtailed rural self-determination. In this context, more sustainable systems, particularly those involving low external inputs, have been recognised as means to nutritional security and food sovereignty. Farming systems that produce a diversity of crops have advantages over monocultures in terms of their capacity to meet the nutritional requirements of communities, their resilience to pest attacks and their relative stability in times of climatic change (Altieri, 2004; Uphoff, 2002a). Furthermore, by reducing market dependency, low-input systems provide farmers with greater scope to determine the kind of farming system they want to create.

The literature on sustainable agriculture highlights remarkable cases of farmers organising self-sufficient and supportive networks for transitions to more sustainable and equitable food systems. The capacity of sustainable farming systems to lessen historical inequalities and overcome some of the imbalances of Green Revolution and neoliberal agrarian development models has made it attractive to Left-wing peasant movements in poor countries, including the Zapatistas in Mexico (de León, 1995; Solís, 1995). Indeed, the transnational peasant movement, La Via Campesina (LVC) identifies sustainable agriculture as crucial to food sovereignty, which is its key organising principle (Desmarais, 2003).

The diffusion of sustainable agriculture through radical social movements has been most pronounced in Latin America. Agroecology has been the chief paradigm used there—a system which builds on the traditional systems of small holders and enhances their productivity through developing more interconnected, high-energy agricultural ecosystems (Altieri, 1984). In Brazil, an agroecology network called the *Assessoria e Servicos a Projetos em Agricultura Alternativa* has established training centres within existing agricultural universities, persuaded small farmers' movements to adopt an agroecology perspective and occupied positions within the government and administration (Altieri and Toledo, 2011). Consequently, hundreds of agroecology initiatives have been established throughout the country, and ecological techniques have become a key part of the agriculture curriculum in many universities. In this case, effective networking within civil society and building consent within existing centres of power have caused meaningful paradigmatic shifts. A more decentralised approach has been adopted in Central America, where, since the 1980s, small holders have established horizontal, peasant-to-peasant networks to exchange knowledge on strategies for sustainable farming and resource conservation (Holt-Gimenez, 2006). While such grassroots initiatives are most advanced in Latin America, they have also been adopted in some parts of Asia. For example, the Peasant Union of Indonesia has established peasant-managed schools for the dissemination of knowledge on ecological farming techniques (Dwyer, 2011).

Within the more socialist countries of Latin America, sustainable agriculture has been part of the state's project of developing national self-sufficiency and resisting American imperialism. Cuba, for example, was forced to adopt a new approach to agriculture in 1989–1990, when trade with the Soviet Union collapsed and the Unites States' trade embargo was tightened. As Rosset and Bourque (2005) outline, the country faced a massive decline in their access to imported fuels, food grains, agricultural inputs, and aid. Given these constraints, the polity embraced a new paradigm, which prioritised use of ecological technologies, redistribution of land, and fair prices for farmers, to incentivise increased per acre productivity. They also placed a new emphasis on local production, with notable developments in urban agriculture. In adopting this new paradigm, Cuba went from being an importer of food in the late eighties to self-sufficient in food by the late nineties (Rosset and Bourque, 2005). Similar policies were adopted by the state in Venezuela after its Bolivarian Revolution (Cunich, 2012).

If one takes these examples in isolation, sustainable farming movements appear as a kind of radical, counter-hegemonic formation. Here are small-holding farmers, subaltern within the hegemony of the corporate food regime, taking a leadership position, building consent for a system based on alternative principles. Indeed, these farmers' initiatives may be considered as *grassroots movements*, defined by Batliwala (2002: 400) as 'movements of, for, and by people most directly affected by the consequences of public policy'. Rather than relying on elite urban outsiders (who lack first-hand experience of agrarian crisis), they are forging viable alternatives to neoliberalism and the corporate food regime *by and for themselves*.

It is in this guise that sustainable agriculture (or 'natural farming', 'permaculture', 'agro-ecology', or whatever other term one prefers) is celebrated by radical environmentalists throughout the world. To give just one prominent example, the environmentalist Vandana Shiva (2016) locates the agroecological practices of the hundreds of millions of small-holding farmers in developing countries at the cornerstone of resistance to the corporate takeover of the planet, which threatens to destroy all life. In an 'epic contest between a destructive and dying outmoded paradigm and a life-enhancing emergent paradigm' (Shiva, 2016: xxii), these farmers are said to be cultivating the relations of care for people and soil that have provided the sustainability of civilisations for millennia.

What these romantic representations overlook, however, is that truly grassroots movements for sustainable agriculture—led by those most marginalised within the corporate food regime and most affected by agrarian crisis—are the exception, not the rule. While these exceptions are extremely valuable as models to follow, fixating on them in isolation risks eliciting confirmation bias regarding the view that sustainable farming is intrinsically counter-hegemonic.

A further problem is that food regime analysis, when it talks of solutions to the crises of the corporate food regime, focuses too much on examples from Latin America and not enough on countries elsewhere in the Global South (Brown, forthcoming). From a Latin American perspective, it might be reasonable to conclude that 'agroecology' and 'food sovereignty' represent nodal points of resistance to the corporate food regime. It would be a mistake, however, to then conclude that this is the natural way in which resistance to corporate agriculture unfolds in developing countries. There are marked differences between the examples of Latin American agroecology cited by food regime analysts and the sustainable agriculture movements that are found throughout much of developing Asia. This should make us question whether the latter can be regarded as examples of counter-hegemonic resistance. Perhaps most crucially, Latin American agroecology has been taken up by peasant movements as the solution to the crisis situation that neoliberalism has introduced into agriculture. In India, by contrast, farmers' movements have largely argued that the solution lies in higher subsidies on chemical inputs and remunerative minimum support prices for agricultural commodities—in other words, a return to the pre-liberalisation,

Green Revolution model of state-supported chemical agriculture. Other Indian farmers' movements, such as Sharad Joshi's *Shetkari Sanghatana* have campaigned *in favour of liberalisation* in agriculture, arguing that it offers commercial farmers the potential to become more profitable (Omvedt, 2005). Whether in favour of liberalisation or a return to Green Revolution policies, Indian farmers' movements do not appear to embody the agroecology or food sovereignty ideals as they are positioned in food regime analysis.[17]

Just as it is not necessary for farmers' movements to support a sustainable agriculture agenda, it is also not necessary for sustainable agriculture to emerge from the revolutionary imaginations of social movements. In India and other countries with high levels of rural poverty, the common reality is for NGOs and development organisations to take the lead. Relatively mainstream members of the development establishment have been promoting sustainable farming systems since as early as the late 1970s. New approaches, such as 'Farming Systems Research' (Gilbert *et al*, 1980; Shaner *et al*, 1981) and 'Agroecosystems Analysis' (Conway, 1985) had criticised top-down rural development interventions, such as the Green Revolution, for lacking a holistic understanding of the ground-level situation of farmers—small farmers in particular. Lacking such understanding, top-down models were seen to promote 'solutions' that were not socially, economically, or ecologically appropriate at the local level. These authors instead advocated interventions that could offer meaningful solutions within particular localities, targeting smaller farmers, with a view of being ecologically, socially, and economically sustainable. These and more recent innovations, such as the 'Sustainable Livelihoods' approach (Scoones, 2015), have become enmeshed within the development establishment since the early 2000s, to the extent that they are almost the new 'common sense' for rural development in the developing world.

Being thus positioned within the development establishment allows the diffusion of sustainable agriculture on a larger scale, yet, as I outline below, it also introduces new sources of influence that subvert sustainable agriculture's potential to challenge hegemonic power relations. Those who lead these kinds of sustainable agriculture initiatives tend to be relatively privileged, compared to the grassroots agroecology movements of Latin America. These individuals are embedded within social relations of power and dependency that are ambiguously positioned in relation to the established hegemonic formation.

---

[17] A possible exception is the Karnataka Rajya Raitha Sangha (KRRS), which has fought against multinational interference in Indian agriculture and in favour of agroecology and bottom-up village development. Yet, as McMichael (2006) acknowledges, the KRRS is predominantly led by capitalist farmers, whose holistic, grassroots rural development model masks deep divisions within villages.

Resources, Political Opportunities, and the Capacity to Mobilise

Forging a strong, well-networked movement for sustainable agriculture requires resources and political opportunities. While in the Latin American context, these resources and opportunities have been consolidated within various 'food sovereignty' movements, in the Indian context, where rural communities are marked by dire poverty and divided by multiple inequalities, the situation is less favourable. Most of India also lacks the extensive history of political empowerment of many Latin American countries, with politics continuing to be highly clientelistic.

A useful point of departure in discussing these limitations is the resource mobilisation and political process schools of social movement studies. The resource mobilisation school, led by Oberschall (1973), noted that any social movement requires a range of resources to function. Resources are here defined broadly and include material resources—including money, labour, and time—but also non-material resources—such as social networks, leadership, enthusiasm, and commitment. For Oberschall (1978), alienated and disenfranchised people are less likely to be in possession of such resources and therefore less capable of forming movements, even if they have strong grievances. From this perspective, movements are likely to consist of people from groups that are already minimally well connected and resourced, who have the capacity to sustain a movement over time. From this perspective, social movements representing under-resourced groups (small farmers, for example) tend to rely on what McCarthy and Zald (1977) term as 'conscience constituents'. These movement participants are typically from relatively privileged backgrounds and have 'discretionary resources', whether money, labour, time, or access to institutions, which they are able to devote to causes that do not necessarily affect them directly. While the resources they provide can be crucial to sustaining mobilisations over time, the participation of conscience constituents can be fickle, being dependent on the whims of conscience.

The 'political process' school has adopted some of these insights and built them into a more comprehensive theory of social movements. In keeping with the resource-mobilisation perspective, they maintain that social movements are most likely to occur where there are existing social networks through which people and resources can be mobilised. Yet, as McAdam (1999) notes, in addition to this, mobilisation requires an opening in the 'structure of political opportunities'. The concept of a political opportunity structure, borrowed from Eisinger (1973), implies the various factors in the political environment that facilitate or discourage people from engaging in political action. This may include a change in legislation that removes the threat of repression, a newfound sense of the mobilising groups' power, or a realisation that established powers are in some way vulnerable.

While political process theorists emphasise the greater likelihood of mobilisation among groups that are well-networked and resourced, they also add that the actions of one group, particularly if successful, may open political opportunities for other, less resourced groups to become involved in the process (Tarrow, 1998: 23–25). In certain situations, therefore, it might be possible for one, relatively well-resourced subaltern group to occupy a position of leadership, providing an opening for those who have historically lacked resources and connections to raise their voices. A key issue here is whether or not groups provide such an opening, or, whether deliberately or for lack of foresight, create 'insular' structures, which exclude subordinate groups from participating.

In studying initiatives for sustainable agriculture, therefore, any tendency to form 'insular' structures and ideologies is not only an issue of equity and representation; it also poses questions of movement efficacy. An initiative that only represents the interests of a single group and does not consciously facilitate political opportunities for other groups can only remain in a relatively subaltern position. Without engaging with other groups, recognising common interests, forging alliances and collaborating, their potential to articulate a new hegemonic formation is significantly curtailed. It is more likely that they will gain concessions from established powers but will not mount a meaningful challenge to the existing hegemonic formation. The initiative will be locked in 'identity politics' and will not develop the necessary connections to become a mass movement.

Mobilising in a Challenging Environment

A flourishing movement for sustainable agriculture or food sovereignty in contemporary India would need to navigate a challenging environment marked by resource constraints and a paucity of political opportunities. The environment is particularly challenging to navigate for the rural poor. Not only must they contend with the limitations inherent in a rural environment, relative to the urban, but the terrain of Indian civil society is highly uneven and privileges the voices of the urban middle class.

Challenges of the rural setting

There is a long intellectual tradition emphasising the constraints on political mobilisation and political consciousness in the countryside. Marx (2005: 131) famously described the peasantry as a 'sack of potatoes'. With this analogy, he implied that while they were relatively homogenous in their way of life, the nature of the peasant economy did not lead to the kind of cooperative relations that could underwrite the formation of a coherent political perspective.

> Each individual peasant family is almost self-sufficient, directly produces most of its consumer needs, and thus acquires its means of life more through an exchange with nature than in intercourse with society ... Insofar as there is merely a local interconnection among these small-holding peasants, and the identity of their interests forms no community, no national bond, and no political organization among them, they do not constitute a class. They are therefore incapable of asserting their class interest in their own name, whether through a parliament or a convention. They cannot represent themselves, they must be represented (Marx, 2005: 130-131).

Taking this statement at face value, subsequent Marxist engagements with the peasantry assumed that they lacked political agency. Independent political action was deemed unlikely, due to the lack of a critical resource: dense social relations. For orthodox Marxists, peasants could, at best, rely on working class vanguards for political representation (e.g., Kautsky, 1988).

When orthodox Marxists did encounter peasant mobilisations, they were dismissive of their politics. Hobsbawm (1973), for example, asserts that while peasant uprisings have been plentiful throughout history and have often been decisive in determining the outcomes of revolutions, the politics of these uprisings tend to be inarticulate. They are based either on local issues or a vague universal peasant identity, lacking political content. In Hobsbawm's view, these limitations were a product of the material constraints on agrarian societies: peasants' relative lack of education and access to communication technologies severely curtails knowledge of the world outside of their immediate transactions. Effectively, they lack the communicative means to see the equivalence between their own struggles and those of other groups, necessary to form powerful national coalitions. Furthermore, Hobsbawm notes the very nature of agriculture constrains peasants' ability to sustain mobilisations. Farmers and agricultural workers are subject to natural cycles. Factory workers can put production on hold while they engage in political activity, but on the farm, the sowing of seeds, harvesting, and maintenance of crops must take place at specific intervals or risk crop failure.

These assumptions have influenced research on agrarian movements. Drawing on a number of historical studies, Oberschall (1973: 140–141) argues that because of their resource deficits, rural populations have historically relied on urban professionals and intellectuals to support mobilisations. From this perspective, conditions are particularly favourable for mobilisation when there are established networks and relationships between urban and rural societies. For example, Oberschall notes that labourers who migrate from villages to towns for employment have historically played a role in disseminating radical ideologies in the countryside that incited rural mobilisations. The view that peasant movements, insofar as they have historical significance, follow a revolutionary agenda originating in the cities was common amongst the major studies on peasant uprisings in the 1960s and 1970s, as was the view that their organisational capacity tends to hinge on the support of urban radicals (e.g., Lewis, 1974; Useem, 1977; Wolf, 1969).

More recent studies of peasant movements have challenged these views. While recognising that there are factors that constrain the political mobilisation capacities of rural populations, contemporary studies show there are other factors that run counter to these. In some respects, the economic and social position of agrarian classes facilitate both movement participation and broad political perspectives. Since the late nineteenth century colonial expansion, international food markets have been densely integrated, making farming communities highly sensitive to the price fluctuations that occur whenever there are shifts in the global political economy (Friedmann, 1978). It is, therefore, in the economic interest of farmers to be aware of global conditions and respond to them politically. Indeed, organised farmers' lobbies have played a strong role in shaping the regulation of food markets to protect them from price fluctuations (Friedmann, 1993).

Claims that rural politics and social mobilisation lack coherence may be the result of historiographical bias. As Patel (2006: 73) notes, bias in mainstream narratives led historians of the twentieth century to neglect the long history of agrarian anti-capitalist movements and the impact these have had on more generalised social movements. This bias was systematically challenged by Ranajit Guha and the subaltern studies school. Focusing on the Indian Independence movement, Guha (1982: 2) boldly argued that the histories of the movement until that point had been elitist, focusing on 'the role of individual leaders or elite organisations and institutions as the main motivating force'. Caught in the spectacle of personalities, such histories overlooked the ways in which common people contributed to the development of Indian nationalism, independently of elites. Such approaches were ill-equipped to explain the diffusion of nationalism throughout the countryside. As a corrective to this elitist historiographic tendency, the Subaltern Studies Group drew attention to a parallel history, constituted by the actions of the toiling masses—a 'politics of the people' (Guha, 1982: 4).

Peasant uprisings were a major focus of the subaltern studies school. Several authors drew attention to the fact that peasants not only formulated their own coherent anti-colonial and anti-landlord agendas in the *absence* of the elite urban nationalists, but frequently, they were in direct *opposition* to them (Arnold, 1982; Chatterjee, 1982; Pandey, 1982). This was chiefly due to the fact that the leaders of the elite sections of the nationalist movement had interests in maintaining the status quo. Nonetheless, as the subaltern studies school was keen to point out, it was the elite upper-strata of the Independence struggle who ultimately dictated the form of the post-colonial state and came to occupy positions of power. This caveat implied a substantial shift in the focus of analysis for the politics of rural movements. Instead of the capacity to form social movements and articulate an ideology, the focus is instead on *hegemonic* power. Closer engagement with the historical record shows peasants are capable of *mobilisation* and forging coherent ideologies. The issue concerns the unequal capacities of different groups to *represent* themselves in a period of intense mobilisation and leverage influence over dominant institutions. The social networks and resources of groups will affect their capacity to represent their struggle as having historical significance and to implement change in a post-mobilisation or post-revolutionary scenario.

These studies demonstrate that we should not underestimate the capacity of agrarian classes to formulate their own political perspectives and assert themselves politically. Yet, it should be noted that relatively privileged groups may be able to devote more time and resources to mobilisation and are often better connected to dominant institutions and the media. As such, when such groups work alongside peasant organisations in rural settings or on agrarian issues, there is always the risk that the former may displace the latter's position in civil society. As the subaltern studies school identified, better-resourced groups from cities, who have greater proximity and connections to national centres of power, may play a hegemonic role in articulating the significance of agrarian struggles, thereby influencing outcomes.

The uneven terrain of Indian civil society

In developing countries, poverty and stark inequalities greatly influence groups' capacities to organise themselves, receive adequate political representation and to effect social change. In India, the issue is particularly complex, as inequalities are multidimensional and have complex historical trajectories. These inequalities have a major impact on the operation of hegemony.

The elite dominance of both formal and informal politics has been an abiding concern of Indian social and political thought. Partha Chatterjee (2004, 2011) identifies how in India (and, indeed, 'most of the world') the bulk of the population is not able to enter the domain of civil society, within which they can be recognised as equal, rights-bearing citizens. Most people lack the means to access basic civic institutions, such as the legal system or the media or to directly petition politicians or bureaucrats. Civil society is, therefore, considered to be a domain for the privileged and educated minority, who have cultivated the credentials and contacts necessary to be 'worthy' of participation in the formal democratic process. As with the 'resource mobilisation' school, Chatterjee emphasises that civil society is not an egalitarian space, as numerous factors limit people's capacity to raise their voice. Caste, community, gender, education, and one's capacity to speak hegemonic discourses are particularly limiting factors.

Lending credence to Chatterjee's claims, a quick tour through the history of Indian civil society highlights an enduring theme of elite dominance. India's Independence movement, though diverse in its participants, was chiefly organised by the Indian National Congress. The Congress was initially formed in 1885 by the British administration, as a vehicle to improve relations with their Indian subjects, but which soon set itself the ambition of independence from British colonial rule. The Congress consisted predominantly of the Indian upper-caste elite: landlords, feudal bureaucrats, and the upper strata of the petit-bourgeoisie (Omvedt, 1993). Drawing on their substantial social and economic resources, they were effectively able to stage mobilisations that applied pressure on the British administration and ultimately achieved self-rule. Not only did they organise the struggle but they articulated the ideologies that would inform the post-colonial state. Responding to the simultaneous movements of more subaltern groups, particularly the peasantry, Congress played a hegemonic role, by making concessions and incorporating some aspects of subaltern groups' demands into their agenda. They thereby preserved the legitimacy of their own right to rule India after Independence was gained in 1947.

Congress sustained the legitimacy of its rule reasonably well in the decades following Independence. The public invested substantial trust in Nehru's project of national development and industrialisation. Insofar as there were social movements during this period, these were mostly 'mass organisations', which were directly linked with the Congress Party and its parliamentary allies, such as the Communist Party of India (CPI) (Mohanty, 1998; Ray and Katzenstein, 2005). Mass organisations mobilised the public to secure electoral support for these parties, while in return, they assumed some direct influence over policy. In other words, these movements organised to initiate reform-from-within, rather than challenge Congress elites' right to rule. Even the CPI tended to suppress expressions of dissent, seeing the hegemony of Congress as part of a broader narrative of historical progress: their leadership was seen to be necessary to the project of industrialisation (Omvedt, 1993).

The contemporary civil society configuration in India emerged in the 1970s as the hegemony of Congress began to break down. The death of Nehru and the dictatorial style of Indira Gandhi undermined Congress' legitimacy within Indian civil society, which in turn created openings for subaltern groups to articulate their concerns independent of party structures (Ray and Katzenstein, 2005). Thus, in the mid-1970s, the social reformer Jayaprakash Narayan was able to launch a massive mobilisation of students and the rural poor, opposing the rule of Indira Gandhi. He believed that only by empowering people at the grassroots would it ever be possible to eradicate poverty and bring about true revolution in the Indian social structure (Samaddar, 2008). Partially in response to these mobilisations, in 1975, Indira Gandhi declared a State of Emergency in which civil liberties and the independent press were suspended and Congress effectively ruled by decree. Yet, during this time, opposition movements continued to build, making renewed demands for democracy and civil liberties (Chandhoke, 2009). When Emergency rule was lifted in 1977, there was a decisive shift in the orientation of social movements. Increasingly, the focus was on empowerment at the grassroots, rather than mobilising support for political parties (Deshpande, 2004; Sethi, 1998). Government support for voluntary organisations, working directly with marginalised communities, increased substantially under the leadership of the Janata Dal, which briefly replaced Congress in the Central government after democracy was reinstated (Kamat, 2002: 9–13).

While these developments provided an opportunity for subaltern groups to form social movements, it would be hasty to celebrate them as signs of a more egalitarian civil society. While there are, indeed, new openings for marginalised groups, their voices are all-too-often drowned out by those of the relatively privileged. A number of authors have argued that the economically, socially, and politically ascendant sections of the Indian middle classes have, in recent decades, constructed an exclusive version of citizenship and activism that displaces the voices of other, more marginalised groups (Anjaria, 2009; Fernandes, 2006; Harriss, 2007). As Deshpande (2003: 129) outlines, the middle classes in India have historically had a 'disproportionate influence on ideological matters'. Under British rule, the middle classes were empowered, both as functionaries of the British Empire and as the leading figures of the Independence movement. Following Independence, the middle classes continued to play a directive role but now within the post-colonial state. Deshpande (2003: 139) suggests that in occupying directive roles as planners and administrators, the middle classes played the historical role of 'articulating hegemony'—that is, they established consent for state-led capitalist development whilst mediating between the claims of various classes and social groups. As liberalisation has dismantled much of the developmental state apparatus on which middle class privilege depended, many members of this class have now diversified, moving into NGOs and other civil society positions, wherein they continue to play the role of 'articulating hegemony'. Thus, while many new opportunities for subaltern voices have emerged in Indian civil society, more privileged groups continue to play the role of speaking *on behalf of* the poor, eclipsing the attempts of subaltern groups to speak for themselves.

The prominence of NGOs

NGOs currently occupy a prominent space in civil society in developing countries. NGOs are not-for-profit entities that are relatively free of government control and typically involved in development work (Vakil, 1997). They are often juxtaposed to community-led social movements, as their agenda is typically set by state and non-state donors, who provide them with funding to implement development projects (Desmarais, 2007).

Like McCarthy and Zald's (1977) 'conscience constituents', NGOs' access to funding provides them with the time and resources necessary to mobilise. They are less bound by the existing political opportunity structure and have the necessary discretionary resources to bring new issues to the table. Yet, the surplus resources these organisations command have led many to argue that their presence may also have negative, distorting effects on civil society. Desmarais (2007: 21–26) notes that NGO projects must align with donor priorities, which may not reflect the immediate needs of communities, let alone their political perspectives. Ultimately, the organisations that are able to market themselves most effectively to donor organisations receive the most global attention, rather than those representing the most worthy causes (Bob, 2002). As these groups gain public prominence, they come to determine what are perceived as the key social and political issues of the times, which may, in turn, affect the activities of social movements (Bartley, 2007). It is thought that the growing NGO presence in civil society may lead to a softening of political demands over time and a tendency towards deploying discourses that are less confrontational (Petras, 1999).

Some have questioned the efficacy of NGOs. Scrambling for funds in a highly competitive international environment encourages NGOs to structure themselves around capturing funding opportunities, rather than delivering effective outcomes (Cooley and Ron, 2002). NGOs often behave more like businesses, responding to material concerns for their own preservation, rather than ideals and values (Sell and Prakash, 2004). Because of this, they are often highly professionalistic and, as such, are seen to provide a platform for educated, middle-class staff, whose voices displace those of more marginalised and disempowered groups (Kamat, 2002; Sethi, 1998). Furthermore, the processes by which NGOs receive and report on funding makes them accountable primarily to their donors and the state—they are not held accountable by their target group or wider society, who have no democratic input into the kind of development work that they do (Gray, Bebbington, and Collison, 2006). For social movements, by contrast, the support of a mass base is far more critical; if leaders become alienated from members, the movement will quickly evaporate. This implies that the proliferation of NGOs can in practice lead to the drowning out of grassroots voices, thereby restraining democratic diversity.

The source of NGOs' funding is another major cause for concern. As I elaborate in more detail below, donors in the current context often have an interest in promoting neoliberal models of development. Because of their dependency on these donors for their survival, NGOs have been labelled by some as amongst the key agents in the promotion of neoliberalism (Petras, 1999). Yet, another key way in which NGOs serve a neoliberal agenda is by providing social services in a manner that complements the rolling back of state-led welfare and development programmes. This, in turn, reinforces the idea that the state is not an effective provider of welfare, justifying its diminishing role (White, 1999). This then may give the state licence to renege on its responsibilities to provide for its citizens (Randeria, 2003).

Importantly, these general tendencies do not represent the experience of every NGO. Not all donors are, like the World Bank, at the helm of the neoliberal hegemonic project. While governments and transnational institutions are important sources of funding, so too are charities, churches, unions, and other groups that are, in many instances, on the left of politics. Furthermore, NGOs respond to more than donor pressures in forging their perspectives. As Ghosh (2009) argues, the politics of NGOs are determined not only through relations with donors but also in their relations with social movements. NGO workers are often recruited from within social movements and may retain ongoing connections with social movement activists, who will continue to inform their ethical perspectives. As such, if NGO workers are in disagreement with the politics of their donor, their relationship may be one of constant negotiation and bargaining (Ghosh, 2009). Thus, while some commentators' characterisation of NGO workers as upwardly mobile, middle class, avaricious, and perhaps even corrupt (e.g., Petras, 1999) will undoubtedly apply to some, it cannot incorporate the diverse and complex reasons that people choose to become involved in NGOs.

Civil society actors have often taken a critical stance towards NGOs. This is particularly the case in India, where a strong tradition of Marxism and Gandhism in civil society has made activists sceptical of the intentions of international donors (Luthra, 2003). This affects the modes in which activists represent themselves and the types of organisations in which they are involved. Patel (2006: 79) observes that:

> Social movements and their members are keen to position themselves as occupants of spaces of unassailable political purity, with NGOs relegated to a far more contaminated realm. Criteria, such as membership, funding and location of offices, are sometimes used to drive a schematic wedge between social movements and NGOs.

La Via Campesina has directly asserted that the strong presence of NGOs in international civil society has been, at best, a mixed blessing. At the 1996 World Food Summit in Rome, LVC representatives challenged the right of NGOs to speak on peasants' behalf and sought an acknowledgement of the unequal power relations between NGOs and people's movements (Desmarais, 2002: 104). LVC has been concerned by the often fraught relations between social movements and

NGOs and has navigated a difficult path trying to forge relations with NGOs that would be beneficial to its own members. It has prioritised forging relations with those NGOs that will provide professional support to an agenda forged by people's movements (Desmarais, 2007). The case of LVC shows that when NGOs' capacity to represent communities is seen as problematic, social movement activists may reassert their greater connection to 'the grassroots' as part of their identity as 'social movements'. This may be a crucial way through which movements gain moral leverage in civil society—despite having less access to resources than their NGO counterparts, social movements may benefit by projecting their own authenticity and grassroots connections to the outside world.

Sustainable Agriculture Initiatives within Webs of Power Relations

While low-external input sustainable farming systems may have benefits for the rural poor, the previous sections show that this is not enough to ensure that people will adopt them. Organisational and material resources need to be in place to promote and facilitate the transition to these alternative farming systems—and these resources need to be mobilised in a challenging environment, hashed with complex power relations. In this section, I go further into the specifics of the kinds of resources required and the kind of power relations that complicate their potential to empower the rural poor and challenge hegemonic configurations.

Sustainable farming systems, while less dependent on external inputs, require substantial knowledge, labour, and good management practices (Uphoff, 2002b: 10). More specifically, their promotion requires knowledge of effective techniques, adequate means of distributing this knowledge, and various forms of support to ensure that decent levels of productivity are maintained and that farmers have access to markets. The initial development of these resources will typically require some level of organisational support. Both the resources and this organisational capacity are typically the domain of the urban middle class.

Knowledge is perhaps the most crucial resource. Advocates of sustainable agriculture often emphasise the importance of local people's ecological knowledge. Across generations, farmers and agricultural labourers who have worked extensively within a particular locality develop a special understanding of its unique features and skills to farm within its ecological constraints. Successful sustainable farming projects thus tend to incorporate this knowledge (Altieri, 2004; Pretty, 2002). Integrating farmers' knowledge into sustainable agriculture projects strengthens the chances of success, by ensuring that farmers participate and become involved in innovation and adapting techniques in response to changing conditions (Pretty, 2002; Uphoff, 2002b). The contemporary environment, however, prevents local knowledge from being able to deliver the desired outcomes on its own. Most practitioners acknowledge the need to begin from the foundations of established local knowledge and practices but to augment these with insights from contemporary ecological science (Altieri and Toledo, 2011; Uphoff, 2002a).

Introducing scientific knowledge into sustainable agriculture projects, however, tends to introduce new power dynamics. Abstract scientific knowledge presupposes an expert in possession of such knowledge and a hierarchy of knowledge within which the expert knows better than communities (Ekins, 1992). Introducing experts, who will generally be educated, urban professionals, creates a potential insider-outsider tension within sustainable agriculture projects. The outcome of such situations largely hinges on how the power imbalance between experts and farmers is negotiated. Pretty (2002) stresses the importance of an ongoing two-way process of 'learning' between farmers and scientists, rather than 'teaching', which implies a hierarchy between those who possess knowledge and those who are ignorant. Furthermore, evidence suggests that in the most successful projects, 'scientific knowledge' does not remain the possession of the elite scientist but is integrated into the collective knowledge of the community (Bunch and Lopez, 2005).

There are several other forms of support which sustainable agriculture initiatives must provide if they are to have an enduring impact on the ground. Biodiverse farmers, in comparison to monoculture farmers, may find it difficult to find access to profitable markets for all of their produce, making it important to develop suitable marketing strategies (Navarrete, 2009; Pretty *et al*, 2002: 253). Policy support is also crucial. Chemical-free farmers can benefit from policies, such as the redirection of subsidies on pesticides towards local development initiatives, support for farmers' cooperatives, provision of market information, and reducing the costs of entry into markets (Pretty *et al*, 2002). Achieving such policies presupposes some form of lobbying or campaigning.

These 'resource needs' of sustainable agriculture suggest that some forms of assistance from 'experts' may be helpful, at least during the initial stages. There is ample evidence to suggest farmers are capable of organising on their own to develop their own knowledge, human resources, and marketing capacity (e.g., Holt-Gimenez, 2006). This notwithstanding, it should be acknowledged that such collaboration is quite unlikely when farmers face substantial material constraints, as is the case for the rural poor in the current situation of agrarian crisis. Mobilisations are more likely to be responses to more immediate concerns, rather than planning for a sustainable future. In contemporary India, external resources are almost always required to get sustainable farming projects off the ground, and at present, these resources tend to come via NGOs, foreign donors, and urban middle-class activists.

This is where the paradox of sustainable agriculture becomes apparent. Part of the appeal of sustainable agriculture is that it appears to offer a subaltern-led alternative to the hegemonic corporate food regime that has given rise to agrarian crisis throughout much of rural India. Yet, to be successful and sustained over time, such initiatives depend on middle class organisers, often operating within an NGO framework. As described previously, both the middle class and NGOs have historically been crucial agents in articulating hegemony and drowning out subaltern voices within civil society. Furthermore, in designing and implementing sustainable farming initiatives, these organisers must forge relations with various actors to give them the legal status, resources, local connections, and moral support necessary to succeed. While different kinds of organisations will engage with different actors, there are five major groups with which the average sustainable agriculture organisation must engage: donor organisations, the state, activist networks, the rural elites, and the rural subaltern. With the exception of the rural subaltern, these groups have all played crucial roles in establishing the hegemony of the corporate food regime.

Donor organisations

Donors are particularly important, as they provide the financial, material, and intellectual resources that make sustainable transitions possible. The necessity for organisational support and technical knowledge on which successful sustainable farming initiatives depend implies a need for trained, usually full-time staff. Having staff presupposes funding for wages, which donor organisations are best positioned to provide in a sustained way.

Having control over the flow of resources gives donors a tremendous amount of influence. In many cases, they are able to decide which projects are able to proceed. As Riddell (2007) demonstrates, a very large proportion of development organisations receive their funding in the form of aid from wealthy countries, bilateral agencies, or multinational financial institutions. Of this funding, approximately three quarters is delivered to organisations to implement projects that donors themselves have formulated, leaving local organisations limited autonomy. For this reason, many scholars have argued that dependency on donors can be problematic, as development organisations may come to structure their activities around maintaining positive relations with donors and securing access to funds, rather than building connections with their target communities (Cooley and Ron, 2002; Gray, Bebbington, and Collison, 2006).

Some of the most prominent donors are institutions whose global influence could only be described as hegemonic. This would include the World Bank, the International Monetary Fund, the United Nations and the United States government, who have all played key roles in the promotion of global neoliberal hegemony. Scholars have argued that by selectively funding development organisations, these institutions can influence them to become agents of neoliberal hegemony, rather than a progressive force for the rural poor (Petras, 1999; Sahoo, 2013). Studies have found that organisations promoting changes that are harmonious with the neoliberal agenda, such as facilitating entrepreneurialism, market integration, and privatisation of commons, are more likely to receive funding than those that politicise development and give political voice to the poor (Gerber and Veuthey, 2010; Ghosh, 2009; Patel, 2006). Indeed, some studies indicate that donor-funded development organisations, rather than empowering the rural poor, serve to pacify them and deflect public attention from more radical, grassroots initiatives (Kamat, 2002).

The state

Another important source of influence over sustainable agriculture organisations is the state. Here, I refer to the state simply as the institutions of government—including the public administration and judicial system. I draw on the Marxist and Gramscian tradition, however, in noting that the state comes under the influence of dominant classes and often acts as an instrument of consolidating those classes' hegemony, even when it carries the legitimating veneer of democratic governance.

Many sustainable agriculture organisations in India have partnered directly with government departments in the shared interest of rural development (Farrington and Bebbington, 1993). In such cases, one may reasonably expect that organisations' activities will align with state interests. Yet, there are less direct means by which state influence can be exerted. In India, the state attempts to ensure its sovereignty by exerting control over the flows of foreign funds to local organisations. The state governs the registration of organisations and, under the Foreign Contributions Regulation Act (FCRA), can cancel the right of organisations to receive foreign funds, if the they are deemed 'political'—which, in practice, may simply mean they are critical of state practices (Jalali, 2008; Kamat, 2002). Both Congress and BJP governments have shown their willingness to use this mechanism to cancel NGO funding. In early 2012, the Congress-led Central government cancelled foreign contributions to a number of NGOs, in response to their members' involvement in anti-nuclear protests, citing suspected 'foreign influence' (Udayakumar, 2012). Since its election in 2014, the Modi government has cancelled funding for an unprecedented number of NGOs, citing both financial irregularities and their 'political' nature. In classical Gramscian fashion, this demonstrates the capacity of the state to shape civil society to conform to its own interests.

Concerns regarding the cancellation of foreign contributions may cause organisations to be cautious in the formulation of development projects. Thus, not only do foreign contributions serve to guide the activities of receiving organisations, the threat of cancelling these contributions by governments can serve to keep the civil society organisations in a passive position, refraining from overt, public challenges to dominant interests.

The state's position on sustainable agriculture in India is somewhat ambiguous. State governments take different approaches, with most still operating within the Green Revolution paradigm. Ironically, this is particularly the case in the states where the most damage has been caused through the use of chemical inputs. In Punjab, where the excessive use of chemicals has been linked to a range of diseases, the two major coalitions in the state engage in a kind of competitive populism to provide the highest subsidies for chemical inputs during elections (Kumar, 2007). The rate of government expenditure on these subsidies in Punjab long ago reached unsustainable levels (Roy and Chattopadhyay, 2009). Since 2010, however, a number of other states that have never had high usage of chemical inputs have announced policies to promote organic agriculture, including Uttarakhand, Sikkim, Karnataka, and Bihar. For the most part, these policies are oriented towards organic certification for sale of produce on national and international markets, which is mostly of benefit to farmers who already have a commercial orientation (Scialabba, 2000). These policies, while potentially providing commercial opportunities for some farmers and addressing some ecological concerns, are aimed more at improving India's exports and thereby consolidating the global corporate food regime, rather than building farmer autonomy in the manner imagined in the food sovereignty paradigm. Much in the manner described by Gramsci, the state selectively absorbs the concerns raised by dissenting voices and incorporates them in a way that continues to empower dominant classes.[18]

Middle-class activist networks

While development studies literature has devoted much scrutiny to the influence of donors and the state, comparatively little attention has been directed towards the influence of national and international activist networks. This may be a significant oversight. Activist networks are important sources of non-financial resources, such as knowledge and expertise, solidarity, and collaborations. Given that research to date has often overlooked their significance, it is worth reflecting on their potential impact in some detail.

In the domain of sustainable agriculture, there are a number of active networks in India through which various organisations share ideas and collaborate. These include the Alliance for a Holistic and Sustainable Agriculture, the Organic

[18] One possible exception to this is the state of Kerala, whose communist-led government introduced policies in 2010 to convert the entire state to chemical-free farming by 2020.

Farming Association of India, the Millet Network of India, and various national anti-GMO campaigns. These networks connect sustainable agriculture activists working throughout the country via annual conferences, e-lists, training workshops and so on. Literature on sustainable agriculture, such as the works of Palekar (n.d.), Alvares (2009), and Dabholkar (2001), can also be considered as a part of these networks, as they provide a basis for the circulation of ideas across vast areas. These networks are quite decisive in setting the tone and focus of the debate on sustainable agriculture in India, and through them, local organisations are connected with international activists and perspectives. Furthermore, while analytically distinct, donors and activist networks are highly interconnected and it is often through networking within activist circles that access to funding is secured.

In the absence of strong local structures to make organisations self-sustaining, connections to activist networks become vital to the development and survival of sustainable agriculture organisations. Organisations may rely on activist networks for moral support and to retain a sense of legitimacy when this is lacking on the ground. Furthermore, those that lack access to large funds through donor organisations may become more reliant on activist networks to provide information, expertise, and contacts to organise activities. Participation in these networks shapes organisations' ideologies, values, strategic orientation, and, ultimately, their practice.

The influence of these activist networks is not always positive, particularly given the urban middle-class background of many of their participants. In his study of the anti-GMO movement in India, Herring (2008) observes that the middle-class activists who led the movement had a class interest in being globally connected and producing articulate critiques that resonate with existing activist discourses. As such, they tended to prioritise the reproduction of transnational anti-GM discourses, rather than remaining faithful to the perspectives of the farmers they claimed to represent. While activists, such as Vandana Shiva were talking of the 'failure of Bt-cotton' for Indian farmers, farmers themselves were not only continuing to buy it in huge quantities but had developed (illegal) cottage industries to reproduce the genetically-modified crop. Herring argues that despite activists' adherence to globally trending anti-GM discourses, farmers' opposition to the new crops was far from clear-cut. Baviskar (2005: 171) makes similar observations of the Indian environment movement, which she claims compensates for its lack of grassroots appeal by mobilising 'dispersed metropolitan support in India and abroad'. There is an ease of association between activists and these dispersed metropolitan supporters, who often have a common class position, language, and worldview. Activists become embedded in these networks in multiple ways, including using literature, joint campaigning, and hosting workshops with participants from other states. Perhaps most decisively, activists may rely on the opinions of those within these networks (whose values resonate with their own) for their personal sense of moral legitimacy. If activists prioritise networking with those who confirm their own worldview, then their discourses may become insular, as they receive less input from those who directly experience the problems that they are representing (in this case, farmers and agricultural labourers). These tendencies towards insularity represent barriers to sustainable agriculture operating in a more dynamic fashion, responding to the immediate needs and perspectives of rural populations.

A connection to such activist networks may seem relatively innocuous at first glance, however, there is a growing literature highlighting the ways that middle-class dominated activist networks are key to the formation of a civil society in India that is hostile to the poor and amenable to a neoliberal development agenda. In the past two decades, there has been a proliferation of civil society organisations in India's cosmopolitan cities that focussed on issues of beautification, urban design, access to public space, and public hygiene (Fernandes, 2006; Harriss, 2007). These initiatives attempt to impose a vision of urban space that reflects the aesthetic sensibilities of the middle classes and their understanding of themselves as cosmopolitan, globalised citizens (Fernandes, 2006; Rajagopal, 2001; Taguchi, 2012). This process is not entirely benign, as the imposition of these aesthetic standards on urban space frequently entails an encroachment on the lives and livelihoods of less privileged groups. For example, a major focus of civil society groups' 'beautification' initiatives in Mumbai has been the removal of hawkers and street vendors to free up pavement space and reduce congestion (Anjaria, 2009; Rajagopal, 2001). Likewise, in Delhi, middle-class environmental activists' attempts to reduce air pollution in the city have almost exclusively targeted small factories and auto-rickshaw drivers, which are major sources of income for the informal working class, while middle-class sources of air pollution, particularly vehicular pollution, are left relatively unchallenged (Baviskar *et al*, 2006; Baviskar, 2011). Taguchi (2012) argues that the methods used by middle-class activists in these kinds of campaigns reflect an elitist attitude. In these urban-cleansing projects, lower classes are understood as 'targets' to be 'educated' on clean, hygienic, and sustainable practices, rather than as fellow citizens with whom they can work together to improve local conditions. In this way, such activism reproduces the position of the middle classes in 'articulating hegemony' (Deshpande, 2003). The poor remain subaltern—they can only be represented as they are blocked from representing themselves.

The available evidence also suggests that middle-class activists and professionals have played a major role in organising and leading rural struggles. These struggles typically do not affect the middle classes directly, being of more direct concern to rural communities. This raises issues regarding representation and how adequately middle-class activists are able to stand in and speak for other social groups. Baviskar (2004), in her study of the movement against the Narmada Dam, found that although the movement mobilised a broad stratum of farmers, adivasis, and environmentalists, the leadership of the movement was distinctly middle class. Yet, more crucially, the organisational structure of the movement was such that the communities who were in the submergence zone did not have any significant input into the movement's decision-making process. This exclusion had a tangible impact on the direction of the movement. The environmental focus of the middle-class activists translated into a continuing focus on outright opposition to the dam. Baviskar (2004) argues that those who would eventually be displaced by the dam could have benefited more from a focus on resettlement.[19] Shah (2010), in her study of the Jharkhand movement comes to similar conclusions. She argues that while the movement claimed to be about protecting the interests of poor adivasis, in practice, the movement was comprised of an elite, middle-class stratum of adivasi activists, with extensive connections to transnational activist groups. These activists had fundamentally different interests to their fellow adivasis in rural areas and relied on simplistic, romanticised representations of tribal culture, which, Shah argues, were ultimately detrimental to the immediate material needs of the local adivasi population.

Rural elites

While sustainable agriculture projects are often formulated in the abstract by donors and activist networks, they take on more specific content when organisations engage with representatives of their target communities. A key issue here is whether activists choose to engage with local elites or the rural poor, as there may be conflict between the perspectives and priorities of these groups. Appealing to local elites has an immediate logistical value, as they are often crucial in bringing about change. Not only are the rural elites typically large land holders (who can therefore make a more significant ecological impact by changing their farming practices), but they are also influential within villages. A change in attitude on the part of elites may, thus, filter through an entire community. With this in mind, organisations often make large land-holding elites their first point of contact.

The influence of elites is pervasive within the rural development sector and has been identified as a major issue by scholars. As Mosse (2001) outlines, elites

[19] Indeed, as Baviskar (2005) outlines, more 'grassroots' movements did campaign for precisely this. Yet, lacking the media connections of the middle-class activists, their voice went largely unheard. The mainstream media, itself dominated by middle class professionals, represented these groups as disruptive and violent, suggesting that they had links to the Maoist movement.

are more likely than the rural poor to recognise opportunities to capture resources in development projects and to have the necessary skills to influence projects' direction to suit their own interests. Given the hegemonic role that wealthy capitalist farmers have played in India over the past five to six decades, it is not surprising that they have the skills necessary to represent their interests as being in the general interest of the community. In an influential paper, Platteau (2004) outlines how, in many cases, elites capture the resources offered by development projects and use them for their own private interests, whilst using the language of the 'common good'. Elites are able to do this because they are generally better equipped with the skills required to deal with external resources and are better able to speak the technical language of development agencies. This 'elite capture' is often perceived to be legitimate by the wider community, given the prevalence of clientelistic political representation in many developing countries, whereby elites act as gatekeepers between the wider community and the benefits of development. Even in cases in which elites are explicitly excluded from participation, the *middle* strata of communities can also dominate and prevent the inclusion of poorer and more marginalised voices in participatory spaces (Funder, 2010). The scramble within communities to seize the benefits offered by sustainable farming projects thus often leaves the poor sidelined.

The rural subaltern

Despite the incentives for sustainable agriculture organisations to work with elites, there is also strategic value in appealing to rural subaltern groups. Depending on local context, this category may include small-holding farmers, landless labourers, women, and members of socially oppressed communities (Dalits, tribal groups, religious minorities, and so on), who have not had a direct role in articulating existing hegemonic relations.

The strategic value of appealing to subaltern groups lies firstly in the fact that they are more likely to be adversely affected by agrarian crisis and thus more open to adopting alternative approaches. Furthermore, if organisations are seen working with the more disadvantaged sections of the community, this is likely to confer on them greater legitimacy—both within activist circles and in the wider public. Indeed, it may be an expectation of donors that organisations work with the poor. Thus, ironically, the relations between organisations and non-local sources of influence, while problematised above, may help to put moral and financial pressure on activists to broker more alliances with the poor and less with rural elites.

Formulating a strategy to implement sustainable agriculture is, thus, always a delicate balancing act between (often competing) social forces. A given organisation cannot be seen in abstraction from these complex social relations. While some of these relations may pull sustainable farming initiatives in directions that strengthen their counter-hegemonic potential (particularly when they push them towards empowerment of the rural poor), others strengthen the influence of currently hegemonic forces which are likely to reinforce the status quo.

Conclusion

In the context of agrarian crisis, sustainable agriculture has the potential to create greater livelihood security for the rural poor. By challenging the dominant position of agri-business and giving the poor greater control over their lives, it also has *counter-hegemonic* potential. Yet, having *potential* does not guarantee that it will be beneficial for the poor nor that it will be implemented in a way that challenges hegemonic interests. That would require, at the very least, that sustainable agriculture aligns with the aspirations of at least a subset of the rural poor, and that projects have adequate resources and political opportunities to be propagated broadly.

As I have outlined in this chapter, in contemporary India, urban middle-class activists and development workers are most likely to have the kind of knowledge and institutional resources necessary to make sustainable agriculture initiatives successful. They are also better positioned to capitalise on political opportunities. Yet, not only are the urban middle classes unlikely candidates for counter-hegemonic leadership in the countryside, to secure the viability of their projects, they also tend to forge links to hegemonic groups. Engaging with donors with links to the neoliberal development establishment provides access to the most secure sources of funding. Engaging with the state, or at least refraining from antagonising the state, ensures project stability. Engaging with middle-class activist networks helps shore up access to resources and garner public legitimacy. Finally, engaging with rural elites is often seen to be crucial to ensuring successful implementation of projects on the ground. While engaging with these forces may allow certain sustainable agricultural techniques to be propagated widely, they undermine claims that sustainable agriculture can be a counter-hegemonic challenge to power relations.

In the case studies that follow, I hope to provide empirical demonstrations of these somewhat theoretical claims. I will show how hegemonic influence is exerted on the ground, often in complex and subtle ways. Even movements that begin with a somewhat radical impulse can either end up reproducing the status quo or be rendered ineffective as a result of their links to the hegemonic forces outlined above. Yet, in presenting these case studies, I recognise the dynamic nature of sustainable farming initiatives and also present some examples of how counter-hegemonic visions are kept alive and how activists can strategise to meaningfully empower the poor—despite attempts to thwart such empowerment by the powers that be.

4

The Kheti Virasat Mission

People's Movement or Agrarian Populism?

'Trentji!' Umendra Dutt's deep voice resounded across his simple apartment in Jaito. '*Idhar Aao!*' [come over here!].

It was late and I was in my pyjamas. I had been staying in Dutt's spare bedroom for the past two weeks while undertaking my research about the Kheti Virasat Mission (KVM), a sustainable agriculture organisation of which he was the Executive Director. Throughout the day, I would spend my time in KVM's office, observing their everyday work, taking note of strategic discussions in the office, accompanying KVM staff on fieldwork trips, and interviewing KVM's activists and affiliated farmers to learn more about their work in introducing natural farming to Punjab—a state which has some of India's most chemically intensive agricultural practices.

When Dutt called me, I closed my laptop, in which I had been writing up notes from the day and hurried into his bedroom.

'*Chappal! Chappal!*' he said, pointing to my slippers, which I always forgot to take off before entering his room.

'*Maaf kijie*' [please forgive me] I said, taking them off.

'*Koi baat nahin, chhoti si baat hai*' [It doesn't matter, it's a small thing] he replied in assurance. 'Would you like to ask some questions?'

I was surprised by the invitation. To date, Dutt had rejected my requests to conduct a formal interview. 'I am not a cow!' he would say, smiling wryly, 'You cannot just milk me for data!' Instead, he advised me to observe KVM's work and my perspective would develop in time. He preferred me to ask him questions informally, when the time was right, and not just approach him with a list of questions.

Not wanting to miss this rare opportunity for a more structured interview, I took out my recorder and asked if he would mind if I recorded the conversation. 'Why are you always recording?' he asked, tutting with irritation.

'Oh, it just helps me keep track of what we've said later. So that I can quote exactly what you've said in my thesis'.

'OK, OK', he said, but I could tell that he wanted me to engage in this research in a more organic way.

From the first question, it was obvious that things would not go by plan. I began by asking about his background and what led him to start an initiative related to ecological agriculture. He closed his eyes and sat in silence for what felt like minutes, as if in deep meditation. His response was characteristically laden with his ecological philosophy.

'See, ecology and agriculture, they are not something different from life. I look upon life with a holistic perspective. It is holistic. No compartmentalisation … '

He went on to tell me briefly about his previous work as a journalist and in right-wing Hindu movements opposing globalisation. Yet, the conversation repeatedly wandered to the deep spiritual and philosophical questions that were his preoccupation. What is agriculture? What is life? What is God, if not nature? At times, I felt as though the intentions of my interview questions were being subverted. When probed on specific details about the organisation's structure and decision-making, Dutt's responses were brief and non-committal. When asked about the goals and objectives of the movement, he did not mention any specific agenda for changing farming practices. Instead, he said KVM's core objective was to make people 'live a life that is part and parcel with a constant ecological consciousness'. When asked about KVM's strategy to achieve this, he said, simply, that KVM tries to 'capture their imagination' and, more luridly, 'awaken their *sanskars*'.[20]

'They should feel one with the ecosystem', he explained, reiterating KVM's objectives. 'They *are* one. They are part and parcel with the entire ecosystem. So when they feel one, they should feel the divine within each and every form of life. In the common consciousness, the earth is not a living form, but soil *is* a living form! We should see that each and every drop of water, every dig of soil—it has some consciousness in it! It is divine. It is part of the divine. If we convey this to the masses, our goal will be achieved'.

Seeing the somewhat lost expression on my face, Dutt suggested we should continue our interview the next day. I felt confused. On the one hand, what Umendra Dutt was saying was quite beautiful, perhaps even inspirational. But it seemed too poetic to give me a concrete point of engagement with KVM's work. I wanted to understand how they engaged with farmers, the challenges of forging relations with donors and the state, their strategies for campaigning, and the way they hoped to usher in a shift to Punjab's unsustainable development model. Instead, I was hearing ideologies related to deep ecology, spirituality, and traditionalism. What I was to learn, however, was that precisely these ideological matters spoke directly to KVM's position in relation to power in Punjab. As will

[20] Though difficult to translate, *sanskars* are a Hindu concept, which could be roughly rendered as 'spiritual virtues', which may be awakened through initiation.

become clear in this chapter, Dutt's epic, often spiritual idiom and refusal to engage in more practical discussions about strategy both reflected the power relations in which KVM was embedded and constrained its capacity for effective action.[21]

In terms of their work on the ground, KVM aims to promote a somewhat radical alternative to Green Revolution agriculture in Punjab and, in the process, to challenge the hegemonic influence of agri-business (particularly multinationals) over its farming systems. Although KVM staff have diverse outlooks, they see themselves as intervening in a multi-faceted 'crisis' situation that is playing out at a regional level. As the epicentre of India's Green Revolution, Punjab has suffered the worst consequences of its side-effects in the form of chemical contamination of the ecosystem (with major implications for human health), water shortages, widening rural inequality, and economic stagnation, all of which have culminated to the point that Punjab's rural society has been described chiefly in terms of 'crisis' since the 1980s. KVM promotes 'natural farming', a method that uses only natural materials available at the village level, as the best solution to the crisis of rural Punjab. Natural farming reduces the ecological impact of synthetic inputs and lowers costs for farmers. Farmers who have applied these techniques with the help of KVM claim to have reduced their debts, giving them greater personal autonomy. Furthermore, by producing food that contains no chemical residues, these methods help alleviate concerns regarding Punjab's health crisis. At a more ideological level, KVM argues that Punjab requires a paradigm shift—away from the reductionist and materialistic perspectives of the Green Revolution and towards a position that is more holistic, spiritual, and within which tradition plays a major role.

As KVM's Executive Director, Umendra Dutt is the central figure framing its perspective. Dutt was born into a middle-class, upper-caste Hindu family in Firozpur district in Southwest Punjab. His professional background was in journalism, rather than farming. Politically, he has had a strong involvement in the right-wing Hindu organisations of the Sangh Pariwar, particularly its anti-globalisation wing, the Swadeshi Jagaran Manch (SJM). In the 1990s, he worked in Delhi as editor of SJM's monthly magazine, Swadeshi Patrika. Through this work, he was exposed to numerous articles on the impact of synthetic pesticides on human health. During his time at Swadeshi Patrika, he was also influenced by a number of activists promoting ecological perspectives on social issues, most

[21] Sections of this chapter appeared in Brown, T. 2014. 'Negotiating the NGO-social movement dichotomy: evidence from Punjab, India.' *Voluntas* 25 (1): 46–66; and Brown, T. 2013. 'Agrarian Crisis in Punjab and 'Natural Farming' as a Response.' *South Asia: Journal of South Asian Studies* 36 (2): 229–242.

notably Anupam Mishra of the Gandhi Peace Foundation. Mishra had gained public attention by espousing India's traditional water-harvesting systems as a means of overcoming water shortages. Dutt was inspired by Mishra's approach, which integrated ecological thinking with an appeal to Indian tradition. After regular visits with this mentor figure, Dutt began to shift his focus from narrowly defined nationalist concerns towards a view that recognised the seriousness of ecological decline.

Image 1: Umendra Dutt

KVM's eloquent and powerful articulations of Punjab's 'crisis' represent a noteworthy ideological challenge to the hegemony of the corporate food regime. KVM represents one of the only organised social forces in the state that are raising awareness of the multiple dimensions of this crisis in a sustained manner. They advocate solutions that appear to address root causes and undermine the basis of corporate control of Punjab's food systems: dependency on external inputs and credit. Nonetheless, the very simplicity of their epic narrative and their rigid adherence to natural farming as the *only* solution to the crisis not only limits the extent of their appeal, it also confines it to a relatively privileged class and gender segment—landholding men. In Punjab's political economy, this group has shared in hegemonic power since the time of the Green Revolution—their empowerment, therefore, does little to alter the status quo. KVM's strategic relations to middle-class audiences, donors, and rural elites go a long way in explaining why they retain both a rhetorical and strategic position whose impact at a state-wide level to date has been relatively marginal.

Punjab History: From Green Revolution to Crisis

Punjab was, in many respects, the epicentre of India's Green Revolution. In 1960, Ludhiana, in central Punjab, was one of the seven districts targeted in the initial phase of the IADP, which had sought to promote Green Revolution technologies in a few key districts spread throughout the country. Uptake of Green Revolution technologies in Ludhiana was pervasive, and it quickly spread to all other districts in the state (Frankel, 1971). A combination of factors made the new technologies particularly well-suited to Punjab. The state's pre-existing, efficient irrigation system, which reached almost every corner of the state, ensured an adequate supply of water to make the new technology gainful (Bhalla and Chadha, 1983), while the consolidation of land holdings and construction of efficient road networks in the wake of partition had aided the development of commercial agriculture (Randhawa, 1974). Crucially, the Punjab government showed a keen interest in facilitating the uptake of the new technologies by building infrastructure and providing farmers with access to credit (Singh and Kohli, 2005). Due to these factors, by the end of the 1960s, most of the farmers in the state were using the technology and wheat production had tripled (Randhawa, 1974). Punjab was established as the 'food bowl of India', a title that it retains to this day: with just 1.5 per cent of India's geographical area, Punjab produces approximately 20 per cent of its wheat and 10 per cent of its rice (Singh, 2009: 262).

From the early stages of its implementation, the Green Revolution brought unequal benefits in Punjab and several unwanted consequences. Although there were increases in productivity across the board, increases were considerably greater for those with larger land holdings (Bhalla and Chadha, 1983). Smaller farmers, with less capital and a greater burden of debt, had restricted access to the new technologies (Frankel, 1971). Frankel (1971) suggests that this unequal distribution of benefits fuelled social divisions and class conflict in Punjab. Similarly, Shiva (1991) suggests that by replacing existing relations of mutual obligation within villages with market dependency, the Green Revolution enabled greater levels of communal conflict to develop. The loss of communal ties meant that in times of hardship, there was less support—farmers were more likely to see each other as competitors than to provide mutual aid (Gill, 2005).

The new technology facilitated greater levels of class mobilisation, particularly from the 1980s onwards. The Green Revolution facilitated a shift to a capitalist mode of production in Punjab and transformed the relations between land holders, tenants, and labourers (Byres, 1981). As Gill and Singhal (1984) note, this created conditions for the contemporary farmers' movements: all farmers, to varying degrees, had been integrated into the market and had a shared interest in maintaining low input costs and high agricultural commodity prices. Strong disillusionment with political parties in the late 1970s led to the formation of independent farmers' movements, which came under the leadership of the Punjab branch of the Bharatiya Kisan Union (BKU). The Green Revolution also saw increased mobilisation of agricultural labourers, who found representation in the communist parties (Gill, 2000).

The Green Revolution did bring considerable short-term economic gains in Punjab, but by the 1980s and particularly the 1990s, the economic situation turned sour. From the early 1990s, growth in yields for Punjab's three major crops of wheat, cotton, and rice began to stagnate, compromising farmers' income (Chand, 1999; Sidhu, 2005). Since 2000, this tendency has worsened, and it appears that yields have reached their maximum. Growth in output has been slow or slightly negative in wheat, stagnant in rice, and negative in cotton (Singh, 2009). As Singh (2009) outlines, this decline in growth has become an economic crisis for farmers, as it coincides with a rise in the capital costs of farming, including chemical inputs and machinery. Furthermore, intensive cropping patterns have depleted soil fertility and, consequently, more fertiliser is necessary to achieve the same yield. Pests have also developed resistance to pesticides, requiring greater quantities of pesticide to be purchased for the same effect. With farmers spending more and earning less, their income has gradually diminished. This has been particularly damaging to farmers with small holdings, who were already in a precarious position. Many small-holding farmers have become highly indebted and subsequently forced to sell their land and an alarmingly high number have committed suicide (Singh and Bhogal, 2014).

In Punjab, perhaps more than elsewhere, the unchecked use of Green Revolution technologies has led to a deterioration of ecological conditions. The water table in Punjab, as throughout much of North India, is at an alarmingly low level (Rodell, Velicogna and Famiglietti, 2009), with many districts registering less than 10 per cent capacity. On top of this, while the water table is being exhausted, excessive surface water irrigation without adequate drainage is also leading to water-logging and salinity (Kulkarni and Shah, 2013). This situation has been attributed to both excessive and unregulated use of groundwater irrigation pumps (Chand, 1999) and the growth of the area of the state under water-thirsty, high-yielding varieties of rice (Singh and Kalra, 2002). While Punjab's governments have been repeatedly advised to promote a shift from rice to corn cultivation in the state, ongoing government subsidies on electricity for tube wells and procurement policies for paddy continue to provide farmers with an incentive to cultivate rice and to drain the water table through tube-well usage (Sarkar and Das, 2014). Equally alarming, there is evidence that the high concentration of agricultural chemicals in surface waterways and in the aquifer is having an adverse impact on human health. In addition to causing environmental damage, such as algal blooms in rivers, there is strong evidence of increased rates of cancers and reproductive health disorders in Punjab (Thakur *et al*, 2008, 2010). Punjab's health issues have become a focus of local media attention and are a source of significant public concern.

This situation has not been experienced in the same way by all classes. The increasingly capital-intensive nature of Punjab agriculture and diminishing returns has had a more severe impact on small and marginal farmers. On a per hectare basis, marginal farmers are three times more indebted than large farmers (Singh, Kaur, and Kingra, 2008). Because of their precarious economic position, these indebted small farmers are over-represented in cases of farmer suicides (Gill, 2005). Declining profitability is making small and marginal farms less viable, and the number of farms in this category has decreased (Singh and Bhogal, 2014). Former small holders now constitute a new 'reserve army of labour' in Punjab, yet the employment opportunities in agriculture have declined dramatically since the 1980s (Sidhu and Singh, 2004), and increasingly, farmers are seeking to employ more readily exploitable migrant labour (Singh, 2012b). The level of local unemployment in rural Punjab has subsequently increased (Singh, Kaur, and Kingra, 2008: 130). Though the media focus has predominantly been on suicides by farmers, suicide rates among agricultural labourers are also increasing (Bharti, 2012). Suicides amongst various rural classes in Punjab, which are above the national average and increasing (Singh, 2009), are frequently taken as a strong indicator of the 'crisis' situation, indicating desperation and a lack of options.

The ecological and health dimensions of the crisis also have uneven effects across classes. As a consequence of the declining water table, only the wealthier farmers can now afford to drill tube wells deep enough to access the ground water, with less affluent farmers being forced to buy water from their wealthier neighbours (Sidhu, 2005). The burden of the health crisis is also felt more profoundly by the lower classes. Singh (2010) has found that due to the poor state of public healthcare facilities in rural Punjab, many small and marginal farmers have avoided treatment, despite a large portion of them suffering from serious illnesses. For farmers in these categories, healthcare expenses were the second most common reason for entering into debt, after production costs.

These changes in material conditions have led to a shift in the way Punjab is perceived and represented. Jodhka (1997) suggests that the 1980s represented a turning point in public perceptions of Punjab: from the model of agrarian prosperity, Punjab came to be seen as a state with multiple problems. Consequently, by the late 1980s 'crisis' had emerged as 'the dominant mode of representing Punjab' (Jodhka, 2006: 1530). The response to this crisis from the state has been slow in coming. The ruling parties have focussed on communal issues and shallow populism rather than addressing the socio–economic dimensions of the crisis (Kumar, 2007; Singh, 2012a). The disconnect between political parties and ground realities had historically facilitated the formation of politically independent farmers' movements (Gill and Singhal, 1984), however, since the 1990s, these movements have become highly fractured, particularly over questions of liberalisation and the nature of their relation to political parties (Gill, 2004; Mukherji, 1998). The 2014 Lok Sabha elections were a telling reflection of the state of contemporary Punjab politics—while the major parties (Congress, BJP and the Shiromani Akali Dal) focussed on the symptoms of Punjab's agrarian crisis (such as youth drug use), the new anti-corruption Aam Aadmi Party (AAP) made substantial (but not yet decisive) gains by focusing on the grassroots concerns described previously (Judge, 2015; Singh, 2014). Thus, while the Green Revolution approach had given focus to development strategies in Punjab in the 1960s and 1970s, the current situation very closely resembles the Gramscian 'crisis of authority': the old way is dying, yet, due to lack of leadership, the new cannot be born. The situation would appear to be ripe for the articulation of radical alternatives that challenge hegemonic configurations.

KVM Origins: Forging an Identity as a 'People's Movement'

In 2000, in response to his observations of the crisis situation in Punjab and his emerging ecological politics, Dutt founded an NGO called Kheti Virasat (Hindi and Punjabi: 'Agricultural Heritage'). The choice of name reflected Dutt's belief that agriculture, viewed holistically, was at the centre of Indian civilisation and that India's loss of its cultural heritage and identity was partly attributable to its abandonment of millennia of tradition when it adopted the Green Revolution. He drew on his political connections within the Sangh Parivar to put together a small team who worked to assist farmers in transitioning to less chemically intensive farming methods. They implemented projects, such as installing rainwater harvesting and vermicomposting units in villages, mostly funded by government grants.

By 2004, however, Dutt became disillusioned. He reached the opinion that the other members of the organisation were not motivated by a genuine desire for change and instead were simply interested in their own career paths. The money flowing through the organisation had attracted the 'wrong kind' of activist, in his view. In February of 2005, he resigned from Kheti Virasat and the next day formed Kheti Virasast Mission. For Dutt, adding the word 'mission' to the title indicated some of the factors behind his split with Kheti Virasat and emphasised his commitment. As he puts it:

> [The change of name] is only to differentiate between us. Because … their running of the organisation, it was for their employment. It was a job. For me it was a mission. That's why. For them it was 'we need money, we need the job security'.

Dutt set up an office for KVM in the small town of Jaito in Faridkot District, having several contacts there from the Sangh Parivar. From the outset, Dutt emphasised that KVM would not follow an NGO approach. In the previous organisation, he felt that this had led to greed and prevented the realisation of key objectives. KVM was to be a *people's movement*, as he explained:

> UD: We call ourselves an environmental action group, or you could say it is a people's movement for environment. We work as a movement, not as an NGO.
>
> TB: What would you say is the difference between working as an NGO and as a movement?
>
> UD: Go and see any NGO. You will find there are two or three persons there deciding their work. Taking projects and doing work. Here the priorities are entirely different. Our priorities are not to take projects, to run for the projects … No. We won't run after projects, and if we take money it is only on our terms. Ok? Secondly, we decide what we want and we do that only. Forget about projects, what any [donor] will say. We do what is urgent, what is important for the movement.

It is worth noting that when Dutt emphatically states that KVM is 'not an NGO at all', he does not refer to its formal, legal status. KVM was, from the outset, registered as a 'trust', which is the typical designation under which NGOs of this size would register in India. Registering as a trust kept KVM's options open for certain commercial ventures, such as establishing a magazine, and was necessary for the receipt of foreign contributions. Dutt's attempt to distance himself from the NGO label, however, was a rejection of the NGO approach as a model of organisation and part of a preoccupation with the corrupting influence of donors. Dutt asserts that the NGO model leads to a commercial approach to public service. He suggests this is closely tied to the culture of contemporary Punjab—wherein all forms of activity, including those in the 'public interest', are seen in terms of their potential to generate money. Dutt and other KVM members also stress that NGO style work ensnares activists in projects, causing them to lose sight of the broader cause. They must pander to the needs of donors—and international donations, according to Dutt, often come with strings attached.

The 'people's movement' label also signalled an unwillingness on the part of KVM to engage directly with the state. In KVM's view, the state government was bought out by the corporate interests behind the Green Revolution. KVM did make efforts to engage with members of all of Punjab's political parties, in the hope of generating an ecological sensitivity that would transcend ideological divisions. However, they did not directly interface with government schemes nor accept government grants, at least during the time of fieldwork.

Dutt's views on NGO funding were echoed by other KVM members. Harmeet, a young man who had become KVM's chief field worker in promoting natural farming in villages, suggested that NGOs are limited in their capacity to raise fundamental issues. They must constantly keep the perspectives and interests of their patrons in mind, whereas a 'movement' like KVM can put forward the most important issues 'without any hesitation, at every platform'. Another KVM activist described NGO workers as subdued, urbane activists, who sit around 'sipping tea', discussing issues and working on their media profile. He preferred a model of activism with 'fighting spirit', which would engage directly with issues and take on vested interests.

At the time of fieldwork in 2009–2010, KVM had maintained a small, committed team of between five to ten core activists, but had a much wider network of supporters. Since that time, it has expanded substantially, as will be described towards the end of this chapter. The office in Jaito remains the centre of activity, but there are small branches in other cities—including Jalandhar, Amritsar, and Chandigarh. To all appearances, KVM has become the most prominent group promoting sustainable agriculture in Punjab. The *Organic Farming Sourcebook* lists KVM as the sole organisation working for this cause in the state (Alvares, 2009). During fieldwork, I did encounter a small number of development NGOs that had included some form of chemical-free farming into their broader development agenda. Dutt is particularly scathing of these organisations, which he sees as opportunistically responding to donor interest in organic farming. He insists they have no long-term commitment to the cause. The presence of these potential 'competitors' to KVM's moral authority on sustainable agriculture, relatively small as they may be, may go some way in explaining KVM's continued need to differentiate itself from them, by emphasising its non-NGO, 'people's movement' identity.

Transitioning from an organisational identity as an NGO to a 'movement' was an attempt to reposition Dutt's relation to donors and also to allow him greater scope to impose his vision on the organisation. Dutt described donor funds not only as a threat to his initiative's autonomy but also a corrupting influence. While not eschewing the receipt of funds altogether, by asserting the newly formed KVM's identity as a 'movement', Dutt was able to attract a kind of activist who was, first and foremost, committed to his vision for Punjab and using funds only towards that end. The move also represented a shift in KVM's public image and an attempt to garner greater legitimacy. As noted in Chapter 3, in India, NGOs are increasingly regarded with suspicion. They are seen as ineffective (Harriss,

2008), with an over-riding concern with making money, rather than promoting positive social change (Jakimow, 2010: 555–557). Shifting to a 'movement' orientation may have been an attempt by Dutt to encourage public confidence in his reformed organisation at a time when NGOs were somewhat on the nose. Yet, if greater public legitimacy was the aim, simply referring to his organisation as a movement, while continuing to operate along NGO lines could easily be met with public scepticism. Referring to itself as a 'movement' put an onus on KVM to do certain things differently—and it has, in ways outlined in the following section (see also Brown, 2014b).

How does a Movement Differ? KVM's Structure and Mode of Engagement

During its first few years, KVM operated in a largely informal manner. This gave it flexibility to develop a broad and active base of followers. Dutt's approach was to involve diverse sections of the community who were concerned by the growing problems related to Punjab's agriculture and to encourage them to make their own (largely voluntary) contributions. Rallies and demonstrations were organised at opportune times, whenever the need arose. Rather than working on 'projects' in the manner of NGOs, KVM initiated a series of 'campaigns' and 'action groups' focussed on generating awareness of specific issues or protesting current developments, such as the commercialisation of genetically-modified crops. For income, KVM relied primarily on donations from individual members and supporters, rather than funding bodies. Dutt asserts that travelling from place to place and requesting donations has allowed KVM to 'build a relationship with the masses'. Making monetary contributions, he suggests, gives people a greater sense that KVM is 'their movement', increasing commitment.

Most individual donations come from KVM's urban supporters. Thus, although KVM's work most directly concerns farming and rural communities, there is a financial need to ensure that the message is amenable to urban Punjabis, upon whom the organisation depends for patronage. Indeed, a surprising number of KVM's activities are urban-centred. Dutt suggests that, to promote a large-scale shift to natural farming, it is necessary to have urban consumers of naturally grown produce and to have people willing to advocate for changes at the policy level. He also strongly believes in raising 'ecological consciousness' amongst the general populace and encouraging people, whatever their profession, to take action in ways most suited to their social position. Developing a multi-level, rural and urban response to Punjab's crisis is a key component of KVM's claim to be a 'people's movement'. To bolster urban support, Dutt himself spends a lot of his time travelling Punjab, giving talks about ecological and agricultural issues at schools, public events, social clubs, and so on. Other KVM activists focus on media and cultural activities, such as writing songs about the environment.

Despite the urban orientation of much of KVM's campaigning, the participation of farmers in the 'movement' is crucial to KVM's claim to represent the grassroots. Of the seven members of KVM's executive committee, four are farmers. Farmers' participation in KVM public events is also crucial to maintain its authenticity in representing farmers' perspectives and concerns. Some of KVM's more passionate natural farmers give talks at workshops and public lectures, detailing their experiences with natural farming and the factors that led them to change their practices. Some also provide tours of their farms to interested persons to demonstrate the viability of their techniques.

A key component of KVM's 'social movement' strategy is forming 'action groups'. These are subgroups within KVM, formed mostly along lines of profession, or on specific issues, in which teams of activists and professionals devise strategies to promote change within specific domains. Dutt has tried to be sensitive to the various social groups that Punjab's ecological crisis concerns and has encouraged them to organise into action groups to facilitate changes in a manner best suited to them. He considers it to be of great strategic importance to involve diverse people, perspectives, and forms of expertise in the movement to make it more dynamic. At the time of fieldwork, there were five main action groups within KVM: the Environmental Health Action Group, Women's Action for Ecology (WAFE), Teachers for Ecological Action, the Vatavaran Panchayat (Hindi: Nature Councils), and KVM's anti-GM campaign.

To bring focus and legitimacy to its endeavours to raise awareness of the health scenario in Punjab, KVM has formed an 'Environmental Health Action Group' (EHAG). The group is composed predominantly of health professionals, who speak at public events and to the media about the links between chemical toxicity and disease. Because doctors are held in high regard, their voices add legitimacy to the claim that Punjab's health scenario is at a crisis point and that some change in approach is necessary. In interviews, prominent EHAG members describe a scenario in which harmful chemicals have become omnipresent in Punjab and in which the individual is helpless to protect themselves from deadly diseases:

> The environment of Punjab is now fully, or I would say, *over*-saturated with persistent organic pollutants in the form of chemical pesticides, weedicides, fungicides—everything. Most of them are very, very toxic. Most of them can be found everywhere. They can be found in soil. They can be found in the subsoil water. They can be found in breast milk. They are in the food chain. All of us are carrying a lot of persistent organic pollutants even in our subcutaneous fat in our body. If you take out my fat you will find it in there. So that is how these chemicals have now entered the food chain and the entire ecology (Prem).

> Sickness has become so dominant in Punjabi houses. Someone is sick. Child is sick, somebody, somebody. There is no family which is free from sickness. One type of sickness or the other … Sickness has become the way of life (Gurdas).

Through statements, such as these, members of EHAG perform the function of 'organic intellectuals'. In contrast to 'traditional intellectuals', who Gramsci (1971: 5–23) describes as atomising people by leveraging social divisions and promoting individualist discourses, 'organic intellectuals' serve to unite the population in solidarity. By articulating the interconnectivity of all life in the ecosystem and the common threat of ubiquitous exposure to harmful chemicals, EHAG's spokespeople show that chemical agriculture should not only concern farmers—it is an issue for all people who live in Punjab. Such a perspective theoretically encourages a collective response. As Gramsci notes, organic intellectuals vary in terms of the extent of their connections to the classes they represent and unify. EHAG intellectuals, being of an urban middle-class background, are not extremely well positioned to express the experience of the rural poor who are most affected by the health crisis. Nonetheless, their efforts give voice and legitimacy to the everyday social realities of living with chronic disease, and their intellectual work provides a language through which this reality can be expressed.

While EHAG relies on the participation of 'experts', other KVM action groups attempt to promote health awareness through more grassroots strategies. The most prominent example of this is Women's Action for Ecology (WAFE). This action group took shape in late 2008 and early 2009, when it was organised largely by an urban activist, Ramneet. Ramneet spent a lot of time working with women in villages near Jaito to promote awareness of the impact of chemical farming on health. KVM activists saw several strategic reasons for developing a role for women within the movement. Primarily, this had to do with a view of the traditional role of women in households and their potential to influence patterns of food consumption, as KVM's chief fieldworker, Harmeet explained:

> Family is basically managed by women. Women have the responsibility to feed the family. And she has the *right* to [decide] how and what she [will] feed. And when women come with KVM, it will give fast progress to us … When the women join us it will naturally go faster and faster.

Ranmeet went further, suggesting that women had a greater sensitivity to their moral obligation to respond to Punjab's crisis situation:

> So we also thought that, you know, it would be a bit hard to encourage a man [by saying] 'What are you growing for your family? Poison?' It would be a bit hard. Because men are a bit hard [laughs] … But, you know, the kind of connection that

> a *mother* has with her child, we can encourage women that 'how can you afford to feed your kids with poison?' If we can build up an atmosphere like this, and if we can start up with just kitchen gardening, from kitchen gardening we can shift to the fields.

Ranmeet was also conscious of the role that women have played in India and elsewhere in the world in rural development:

> Another thing is that if we see the development work that is being done in various parts of the world, we have seen that wherever the development work has been done it has mostly been done by women ... You know women, the inherent nature of give and care, their energy can certainly be diverted to such work.

Ramneet had organised meetings and festivals in villages to promote awareness amongst women of the relationship between disease and pesticide usage. Rather than simply quoting statistics to the women, she would engage in dialogue with them and encourage women of different generations to discuss changes in the health of communities over the years. Subsequently, women made rudimentary attempts to document the extent of cancer and other diseases within their communities and compare current rates of cancer to those in times past. Furthermore, at festivals, older women were involved in reviving old recipes for pre-Green Revolution crops, particularly drought-tolerant millets, which KVM feels will eventually make a more sustainable alternative to rice for Punjab. Initiatives such as this might not have led to the kind of measurable outcomes desired by international donors but, at least in theory, may lead to a dramatic cultural shifts in the ways in which Punjab's agrarian issues were perceived.

WAFE was particularly influential in the village of Bhotna, where Ramneet made an important local contact, Akashdeep. Akashdeep's husband was already practising natural farming with KVM's help and through this she had already been exposed to KVM's perspectives. She was convinced of the moral imperative that women should not feed chemically-contaminated food to their families. To spread this message throughout the village, Ramneet and Akashdeep began to organise meetings with women to share information and teach them how they could begin to grow their own chemical-free vegetables in their courtyards and kitchen gardens. In the beginning, there were only a handful of women who came to meetings, but by the time of my fieldwork, Akashdeep claimed that approximately thirty-five women in the village had taken up home gardening, and one woman had also started natural farming on a small acreage. She also claimed that some of the women had been able to convince their husbands to convert a small part of their landholdings to natural farming to meet the family's subsistence requirements. Akashdeep and other women in Bhotna were helping their husbands in preparing biological inputs for natural farming. It was acknowledged, however, that there were limitations to this, as many women in the village were not given the freedom to attend meetings and had various other 'constraints' in their home environment.

In late 2009, Teachers for Ecological Action (TEA) was formalised as a new KVM action group. KVM had previously developed a support base of up to forty teachers, who had been involved in promoting ecological perspectives in schools through Eco-Clubs[22] as well as poetry and essay-writing competitions. TEA was formed as an attempt to focus this activity. The main organiser of this new action group, Prabhir, suggested in interview that teachers are particularly well positioned to help expand awareness of ecological issues in Punjab. He suggests this is because they are held in high regard in communities, often enjoying the status of 'Gurus'. Furthermore, they have regular exposure to both students and parents alike, in both rural and urban areas, and perform the important function of engaging with ideas and bringing them into the lives of communities. Crucially, Prabhir argues that teachers have spare time to devote to social service which those in other professions lack. TEA was still developing during the fieldwork period; however, they were already devising strategies to bring ecological perspectives into the lives of students, both inside and outside of school. Prabhir and others also showed a keen interest in campaigning to influence the school syllabus to make it more sensitive to the scope of the ecological crisis in Punjab, though it was acknowledged that this would be a challenging task that would require a high level of organisation.

The action group through which KVM behaved most like a 'social movement' in the traditional sense of the term was its anti-GM campaign. Dutt had been involved in the movement against GM since his days in SJM. During that time, he came to see GM crops as part of an American imperialist conspiracy to control India's food systems. KVM has been able to rally farmers around this cause, particularly those who claim to have suffered as a result of the failure of Bt-cotton. KVM has held numerous rallies protesting the commercial release of GM crops and was particularly prolific during public debates on the commercial release of Bt-brinjal. Despite mobilising farmers for its anti-Bt-brinjal campaign, however, KVM's public statements on the issue mostly focused on the potential of the new vegetable crop to further contribute to the 'poisoning of the food supply'. Rather than emphasising farmers' concerns for increasing costs of production, KVM

[22] Eco-clubs were a part of the National Green Corps initiative of the Ministry of Environment and Forests of the Government of India, which commenced in 2001. Eco-clubs are now operating in over ninety thousand schools throughout India. The main objectives are to raise student awareness of ecological issues and actively involve them in conservation efforts (Roberts, 2009).

fixated on existing middle class concerns with pesticides and Punjab's health crisis. It was left to farmers' movements to speak more directly to the economic impact that the Monsanto-owned seeds might have on farmers.

Image 2: A poster for KVM's campaign against bt-brinjal

The Chiranjivi Gram Abhiyan

It is noteworthy that, with the possible exception of WAFE, KVM's initiatives that were most central to its claims to be a 'social movement' predominantly revolved around the interests and perspectives of the urban middle class, on whom it depended for moral and financial support. Even its anti-GM campaigns, while certainly relevant to farmers who feared the rising costs of seeds, were mostly oriented towards urban consumer concerns with the poisoning of the food supply. In its engagements with villagers, however, KVM tended to operate more along the lines of a traditional NGO. This was most evident in the operations of its natural farming campaign, the 'Chiranjivi Gram Abhiyan'.

In the early stages, KVM's attempts to promote natural farming in villages entailed working with farmers who had an interest in experimenting with these techniques and providing them with basic support and information. In 2008, they began to work in a more focussed way, starting on a sponsored project, in collaboration with the Centre for Sustainable Agriculture (CSA), a Hyderabad-based NGO. The project was called the Chiranjivi Gram Abhiyan (CGA), which roughly translates to 'Sustainable Villages Campaign'. Its aim was to completely convert five Punjabi villages to natural farming techniques and thus provide model villages for sustainable development in Punjab. To this end, CSA assisted in mobilising experts for focussed training programmes and KVM provided local connections and permanent staff. KVM selected the villages to be targeted. Selection of villages was made on the basis of proximity to KVM's office in Jaito, and certain strategic considerations (focusing, for example, on villages with high rates of cancer). The villages targeted in the CGA also became the focus of several of KVM's other initiatives, such as WAFE. The project was funded by the Sir Dorabji Tata Trust, the charitable foundation of a famous Indian capitalist dynasty.

The day-to-day operations of the project were coordinated by Harmeet. Harmeet had been an activist with KVM for several years before beginning work on the project. Though his formal education was in politics and Hindi literature, he had undergone several training programmes in natural farming with KVM and developed considerable knowledge of effective techniques. When not engaged in other campaign work, Harmeet would travel each day to one of the five villages and ensure that techniques were being implemented correctly and that farmers were not experiencing any difficulty. In each village, there was a local coordinator who had received intensive training, whose task was to maintain more regular contact with the practicing natural farmers, particularly during key phases of the crop cycle. Harmeet and the five local contacts were each paid a modest monthly salary for their work. The CGA also organised a series of events to help educate farmers about natural farming. At these events, experts and practitioners from around India gave lectures and workshops on topics, such as mixed cropping and the preparation of bio-inputs. Many of those who ran these workshops either worked for or were affiliated with CSA. Support from these professionals demonstrates how the CGA project enabled KVM to more effectively mobilise expertise.

By grafting the CGA onto its agenda, KVM effectively operated according to both social movement and NGO modes in different contexts, behaving as what Hasenfeld and Gidron (2005) term a 'hybrid organisation'. Harmeet viewed this dual strategy as being appropriate to the aims of KVM. Regarding the project work, he asserts that KVM has a 'duty' to educate farmers on natural farming techniques, given that they have been so actively promoting them. This requires the kind of resources and expertise that an NGO structure provides. Conversely, he suggests that KVM's identity as a 'social movement' provides them with legitimacy that mainstream NGOs lack. He cites the proliferation of NGOs in Punjab in response to funding opportunities as evidence of their inferior moral status. Since KVM has been advocating these techniques for many years on a meagre budget and since they still devote so much of their time to campaigning, their motives appear less questionable. Harmeet does not indicate any tension within KVM in having two different strategic approaches and suggests that they are simply two different modes of action. In the village context, he says, when KVM is engaging with farmers to change their views on farming, they are working in a 'movement' mode; when they educate and try to build the capacities of farmers, they work in a 'project' (NGO) mode.

Image 3: A natural farm in Faridkot District, developed through support from the Chiranjivi Gram Abhiyan

The class dimensions of this 'hybrid' organisational model are striking. If KVM is a 'people's movement', it is worth asking who among the 'people' it represents as a constituency and who are merely the targets of its interventions. While NGOs have been criticised in the past for the dominance of urban middle-class professionals within the organisation, the substance of KVM's repositioning itself as a 'people's movement' did not do much to reverse this—in fact, in some ways, it reinforced an urban bias. KVM behaved like a social movement in its engagement with its urban middle-class supporters (which in turn informed its ethos) but retained NGO-style paternalism in most of its engagements with rural communities.

'Our Agricultural Heritage': Natural Farming and Agrarian Populism

KVM's organisational structure and relations to its urban middle-class supporters clearly suggest reasons to be sceptical of its claims to represent rural concerns.

Deeper investigation into the discourse that KVM promotes reveals further reason for scepticism. Not only did this discourse reflect and reproduce an urban perspective on the issues (which often rendered it incapable of generating meaningful change in rural settings), it also contained traces of a kind of 'agrarian populism' that privileged the position of relatively well-off groups within KVM's target communities (Brown, 2013).

The very concept of 'natural farming', which informs most of KVM's interventions in the villages, provides a useful starting point for exploring their discursive position. KVM does draw some philosophical inspiration from the works of the Japanese pioneer of natural farming, Masanobu Fukuoka (Fukuoka, 1978). Yet, in terms of actual farming practices, KVM's version of natural farming is more inspired by the work of Subhash Palekar of Maharashtra, whose books on 'spiritual farming' have become influential in sustainable farming circles throughout North India (Palekar, n.d.). Palekar's 'spiritual farming' and KVM's 'natural farming' are both distinguished from other farming systems by the fact that they involve neither the use of synthetic chemicals *nor* the purchase of any external inputs from outside of the village. This distinguishes it from 'organic farming', which may involve the purchase of organic inputs from the market. Harmeet described the difference as follows:

> [Natural farmers] never use any inputs from the market. But in organic farming, farmers have to use inputs from the market, like biofertilisers, biopesticides, *etcetera*. But in natural farming, farmers prepare their own biopesticides and biofertilisers with the material which is available in their home, or farmyard, or in their field, or nearby the village. So this is the biggest difference. According to this difference, organic farming is basically market-oriented farming … The products of organic farming … they are very costly … Only some people can afford. So basically organic farming is a farming with the landlords, but not for marginal farmers.

Additionally, KVM activists point out that organic farming is bound up in systems of third-party certification. Farms are evaluated by independent organisations to verify that there are no traces of chemical inputs on the farm, and issue certificates, which provide consumers with certainty that the organic food they are purchasing contains no chemical residues. The process requires the submission of cumbersome amounts of paperwork and is expensive, making it prohibitive for small farmers (Khosla, 2006). Thus, KVM argues that 'organic farming' can only exacerbate the current crisis situation for farmers, placing additional costs on farmers to purchase both organic inputs and certificates. Natural farming, by contrast, does not involve excessive costs.

Image 4: A farmers' shed, displaying recipes for bio-inputs using locally available materials

KVM activists argue that natural farming is the ideal response to the crisis situation in Punjab. Since natural farming does not involve the use of synthetic chemicals, it addresses the concerns about ecological toxicity and public health. Furthermore, costs involved in natural farming are much less than those in chemical farming, reducing the need for farmers to enter into debt. For some farmers affiliated with KVM, these ideas had a strong resonance. Jagmohan, for example, had spent decades campaigning with Leftist parties and farmers' movements for subsidies and support prices and ultimately became frustrated, perceiving that the gains made could never keep up with the growing pressures. Wanting an alternative, he sought out NGOs, but found they were all promoting organic farming and new forms of commercial ventures, which he was not interested in. When he came upon KVM and their natural farming model, he saw a genuine way to break out of the cycle of high expenses and indebtedness that commercial agriculture entailed. Indeed, other practising natural farmers told in interviews that all the NGOs in Punjab only promote organic farming, whereas KVM was the only group promoting a model that offered a real alternative.

Although appealing to this small fringe of farmers looking for a radical departure from the status quo, one wonders whether KVM's rigid adherence to 'natural farming' and staunch opposition to 'organic farming' was strategically sound. Notwithstanding their valid arguments against organic certification and the costs of organic inputs, the fact remains that Punjab is a state in which agriculture has been highly commercialised for several decades. The challenges of engaging with the organics market for small-holding farmers in regions where agriculture is predominantly subsistence oriented might not have been so great for Punjab's market-savvy middle peasantry. The rigid adherence to only one model of sustainable farming practice may have appealed to KVM's environmentalist, Gandhian and Hindu nationalist urban supporters. Yet, this same rigidity may have limited KVM's potential to develop a broad following amongst farmers of various categories in rural Punjab.

In their attempts to promote a shift in agricultural practices in the state, KVM has developed its own interpretation of the 'crisis' situation in Punjab. Although engaging with the pressures of commercial farming and indebtedness, KVM places particular emphasis on the health dimensions of the 'crisis' situation in Punjab. Harmeet said that he considers health to be the key factor that motivates farmers to change their practices. When working in villages, he encourages farmers to think of the toxins to which they are exposing their families by allowing them to consume food grown with chemical inputs. In his public lectures, Dutt talks predominantly about Punjab's public health situation, focusing on how the accumulation of chemicals in the ecosystem is causing a decline in the health and vitality of the people and ecology of Punjab.

Perhaps more controversial than KVM's focus on health, however, is Dutt's frequent claim that the situation in Punjab is essentially a 'civilisational crisis'. Presumably drawing on his background in the Sangh Parivar, Dutt argues that India's religious and farming traditions have been gradually eroded by the Green Revolution and that natural farming represents a way for Indian farmers to 'reclaim their glory'. The nuances of this perspective must be carefully unpacked. On one level, it relates to the notion of sovereignty. Dutt and other KVM activists argue that the Green Revolution was an imperialist intervention into Punjab, which was designed to benefit multinational seed companies. Similarly, they claim that the new push to grow genetically-modified crops in Punjab is also part of Monsanto's agenda to control Indian agriculture. They suggest that, in order for farmers to avoid this foreign threat, they should embrace natural farming and have complete control over what they introduce into their farming system. Yet, beyond simply having autonomy from external forces, Dutt and KVM advocate that farmers should become more connected with their 'traditions'. Indeed, the very notion of 'agricultural heritage' (the literal translation of *kheti virasat*) suggests that what KVM is aiming at is a kind of revival or reclaiming of the past. When asked about the name of his organisation, Dutt explained:

> Ours is a country which has developed agricultural implements about ten thousand years back. And agriculture is a part of our heritage. So I think that to revive our glory it is very important to revive our agricultural heritage, to *know* our agricultural heritage. Because these days we have somehow forgot the agricultural heritage of ten thousand years, and we are following the agricultural science which is only a hundred years old.

Despite this rhetoric, is worth noting that the farming methods that KVM promotes are often not particularly 'traditional'. The methods that they promote involve a system of mixed cropping that has been calculated to optimise nutrient cycles and the application of bio-inputs from cow dung and cow urine which encourage the growth of nitrogen-fixing bacteria. It thus draws in meaningful ways on modern scientific knowledge. The reclaiming of heritage is more a matter of ideology, disposition, and attitude. A strong component of Dutt's emphasis in this respect is recognising the 'spiritual' dimension of farming, which he claims to be an inherent part of India's religious traditions. Dutt often states that 'God is omnipresent in all of nature and therefore any act that harms nature is violence against God'. As such, he suggests that the ultimate goal of KVM is to awaken the people of Punjab to the divine presence in nature, by virtue of which they will cease to commit destructive acts.

Of the natural farmers interviewed for this research, only a small number expressed concerns for sovereignty or spirituality as a major factor that motivated them to adopt natural farming. Most of them were motivated by concerns about health and some by economic concerns (indebtedness, etc.). The one farmer who did express an interest in spirituality was less interested in reconnecting with God in nature—much less 'tradition'. He was more interested in the joy that natural farming brings and the contrast between the violence of chemical farming and the non-violence of natural farming. Even for this farmer, however, the spiritual dimension of natural farming was not the initial reason for his 'conversion'. As such, this focus on spirituality, sovereignty, and the 'civilisational crisis' of Punjab did not appear to be a useful strategy for motivating farmers to reduce their use of chemicals, though it may provide some incentive for them to continue natural farming once the transition has been made. The 'civilisational crisis' perspective is more closely related to the fact that (a) KVM activists believe that after awakening to the agenda of multinationals, the people will take the movement to the next level; (b) KVM's self-representation as a spiritually and politically conscious movement provides legitimacy, particularly among its urban supporters, who give it financial backing; and (c) it gives legitimacy to a strategy of 'agrarian populism', discussed below.

KVM's discourse does not place a great emphasis on economic issues. Certainly, KVM does state that, through reducing their dependence on chemicals, farmers can reduce their expenses and overcome their debts, and a section of the farmers working with KVM have adopted natural farming techniques predominantly for this reason. Yet it is clear that KVM does not see the economic argument as their key strength, due to the fact that adopting natural farming usually involves a reduction in yields, at least over the short term. A number of KVM members and affiliated farmers mentioned that natural farming would be more economically viable than chemical farming if it were not for government subsidies on chemical inputs. Several farmers and activists claimed that if instead of this subsidy, farmers were given a direct cash transfer on a per acre basis, then large numbers of farmers would convert to natural farming. One of KVM's key policy demands is a shift to direct cash transfers to farmers, but KVM saw this as primarily being an issue to be taken up by farmers' movements. Furthermore, KVM's leaders were aware that farmers' movements had made this demand in the past and not been successful (in their view, due to the undue influence of chemical companies on policy-makers) and, hence, were not optimistic about prospects for reform.

Agricultural labourers were strikingly absent in KVM's discourses on the benefits of natural farming and the 'crisis' of Punjab. Some of KVM's supporters noted that labourers stand to benefit from chemical-free agriculture, as it is more labour intensive, increasing employment opportunities and thus putting upward pressure on wages. Some also suggested that agricultural labourers benefit from natural farming, as they are not forced to apply chemical sprays, which are unpleasant and damaging to their health. Conceivably, KVM could assist labourers in agitating for safer working conditions by demanding a limit on injudicious chemical use. Yet KVM has deliberately avoided interactions with agricultural labourers, due to concerns that this would alienate farmers, whom they consider to be their key constituency. Uttam, a former KVM activist with a keen sense of social justice, ultimately left the organisation, partly over what he saw as a disinterest in agricultural labourers within the organisation. He claims that while travelling to villages, activists would never speak with labouring families and would associate exclusively with landowners. His attempts to hold meetings with labourers while he was working with KVM were met with discomfort by other KVM activists. Uttam was also concerned that natural farming would increase the price of food, which would adversely affect labouring families. KVM's ambivalence about labourers casts doubt on its proclaimed strategy of working with all people whom the crisis of Punjab affects and of developing a broad and dynamic intervention into the crisis situation.

KVM's (lack of) engagement with subaltern groups, such as agricultural labourers, speaks volumes about its discursive position. KVM prioritises being 'non-disruptive' within rural communities, which, in practice, means not acting to antagonise the group whom they see as their core constituents—landholding farmers (who, in Punjab, are overwhelmingly male). The discourse resembles that described by Brass (1997), and elaborated by Nanda (2003), as 'new agrarian populism'. Nanda argues that the populist rhetoric adopted by the Hindu right has taken root amongst sections of the middle peasantry who were economically empowered by the Green Revolution, yet have been threatened by the breakdown of traditional caste and gender relations. Hindu nationalist discourses, which naturalise and romanticise traditional hierarchies, serve to neutralise potential assertions from subaltern groups. KVM's rhetoric, particularly regarding the 'civilisational crisis', certainly resonates with this view. Its approach to promoting natural farming has in some ways reproduced traditional hierarchies. KVM's refusal to engage with agricultural labourers ultimately contributes towards labour's further marginalisation and the empowerment of the labour-employing middle to large peasantry, who have shared hegemonic power within Punjab's political economy since the time of the Green Revolution (Byres, 1981).

KVM's engagements with women can be read in a similar light. Although WAFE seeks to involve rural women in KVM's 'movement', it does so only in kitchen gardening and reviving traditional recipes. Failing to provide more genuine empowerment to women or to challenge their subordinate position in Punjabi society again panders to the interests of male land holders, who benefit from women's exploited domestic labour. The gendered division of labour is reproduced and, in some ways, strengthened by KVM's promotion of discourses that celebrate rural tradition, including traditional family roles.

Embracing romantic notions of village life, ecological harmony and the spiritual emancipation represented by farming have not been a particularly effective way of encouraging farmers to reduce their use of chemicals or engage in more sustainable farming practices. Yet, romanticising rural tradition undoubtedly serves to gloss over the highly unjust caste, class, and gender relations that have defined (and continue to define) rural India. It is worth noting, however, that there are limits to the extent to which KVM's discourse can be described as 'populist'. Rather than exclusively empowering the rural elite (large land holders), KVM is most interested in engaging with small-holding farmers—and, indeed, most farmers affiliated with KVM at the time of research had less than ten acres.[23] This may have been because these farmers had been most affected by agrarian crisis and therefore most open to alternatives. It may also have been a matter of maintaining an image as a 'people's movement' with a sensibility for social justice. The influence of donors

[23] Note that while ten acres would be considered a large land holding elsewhere in India, in Punjab, it would be close to an average-sized holding.

and activist communities is likely to have played some role in encouraging KVM to distance itself from large land holders, undermining, to a certain extent, the legitimacy of these traditional elites.

A Movement No More? KVM in Transition

With its traditionalist discourse and strategies for engaging with rural communities, KVM tended to reproduce the hegemonic balance of power at the local level, rather than challenging it. Furthermore, through its major linkages to urban middle-class groups in its various campaigns, it tended to adopt their views in forging its perspective on Punjab's crisis situation—something which reduced the extent of their impact on the ground. In forging an identity as a social movement, however, and distancing itself from donors and the state, KVM at least seemed to resist the hegemonic neoliberal rhetoric that those entities tend to encourage. Yet, at the time of fieldwork, KVM was undergoing a significant restructure—one that would bring it more in line with a traditional NGO model and normalise relations with donors. The processes by which this occurred are illuminating in demonstrating how donor influence persists, even when activists within sustainable agriculture organisations vehemently resist it.

The CGA, mentioned previously, already represented an NGO-style component to KVM's engagement with farmers: it was essentially NGO 'project work'. Yet, during the fieldwork period, it had become apparent to KVM's leaders that difficulties in implementing the CGA project suggested that the entire structure of the organisation required rethinking. KVM was in the process of considering ways to provide more formalised structures in the organisation so that it could better implement its central agenda—the promotion of natural farming.

The challenge of natural farming in Punjab

The CGA had mixed success. During the research period there were between ten and twenty-five farmers practicing natural farming in each of the five targeted villages—significantly below targets set by CSA for that phase in the project. Furthermore, most farmers practising natural farming as part of the CGA were only doing so on a small scale. These modest results indicate some of the challenges associated with promoting natural farming in Punjab but also some of the limitations to the approach KVM had taken.

As noted previously, although farmers in the CGA were motivated by a variety of factors in choosing to adopt natural farming, the overriding reason that farmers cited in interviews was health concerns. Knowledge of Punjab's health issues is quite pervasive. Through interactions with KVM, this awareness could be translated into new forms of practice. As farmers attend KVM lectures or workshops they are able to establish connections between health problems and farming practices and consider possible alternatives. Although this had inspired a few farmers to convert all of their land holdings to natural farming, the most common pattern observed in all villages of the CGA was for farmers to convert 1–2 acres of their land holdings to natural farming methods and continue to farm with chemicals on the remaining acreage. Land holders and their families would consume food grown without chemicals, while continuing to sell chemically grown produce on the market. Given that they were motivated by concerns for their families' health, there was no incentive to change their farming practice, beyond producing chemical-free food to meet their own domestic requirements. Most farmers interviewed suggested that they would cultivate more of their land holdings using natural techniques, if they found that natural farming is equally productive. It seems they have not. During the research period, the project had been running for over two years, and most farmers were still only cultivating one to two acres without chemicals.

Despite the claims made by KVM, farmers, with a few exceptions, indicated that using natural-farming techniques had failed to deliver comparable yields to chemical farming. Part of this may be attributable to the techniques themselves. It may have been unrealistic to expect that without synthetic inputs, farmers could still achieve the high yields Punjab's farmers had produced following the Green Revolution. It was also apparent, however, that the natural-farming techniques were often not being implemented completely, which points to organisational problems within KVM. It was clear from fieldwork experience that local coordinators for each village were providing incomplete and inadequate advice to farmers. Harmeet lamented that these coordinators lacked commitment and said that it was difficult to find people at the village level with an interest in this kind of work who possessed the appropriate skills. Most of those skilled in performing a supportive role for farmers preferred to seek white collar work in nearby cities, he claimed.

After more than two years of being involved in the project, some farmers had become disgruntled with KVM and the idea of converting to natural farming. During an interview, one of KVM's more committed farmers, Ashish, whose five-acre plot had been used as a showcase of biodiversity by KVM, complained of the lack of ongoing input:

> TB: Do you feel like you've received enough support from KVM?
>
> AM: No sir, very little. The only support we are getting from KVM is knowledge. No support to sell produce. No other support.
>
> TB: Has the [CGA] brought any benefit?

> AM: No. We worked for some time under this scheme, but there was no benefit. I suggested to them that instead of going with a large number of farmers, that you choose five farmers from different parts of Punjab, and let these farmers experiment with natural farming. And these farmers should receive all kinds of support from KVM. Financial support, manual labour support. Marketing support also. And if it is successful, then other farmers will follow that.

Ashish's farm had been running at a loss and, in particular, he had struggled to find appropriate marketing avenues for his produce. His comments above reflect an expectation on the part of farmers that despite their lack of funds, KVM should be behaving more like an NGO and giving comprehensive support to farmers as they make their transition to natural farming. Other interviewees commented that in KVM's general meetings, other farmers had started to express their disillusionment. Some members were concerned that farmers would soon give up and go back to chemical farming, which could be disastrous for KVM's image. The experiences with the CGA and the concerns they have provoked were driving KVM to once again rethink its strategy and organisational identity.

Towards a 'professional' approach

Its limited success in the villages had led KVM to an internal crisis. During the fieldwork period, there was a broad consensus within the organisation that it was failing to live up to its mandate to show the viability of natural farming in Punjab. The need for a new approach had become apparent in general meetings, where farmers expressed their discontent. Dutt made it clear to me that KVM was in a state of transition and that it needed to reorganise its resources to become a more 'professional' organisation. This call to increase KVM's professionalism was not new. Since CSA became involved with KVM in 2008 (sponsoring the CGA), addressing KVM's apparent lack of professionalism had been a priority. CSA had successfully implemented sustainable farming projects on a large scale and felt that a different organisational structure was necessary for KVM to have similar success in the field. As such, one CSA member, Sangeeta, was called on to help bring about reforms. Initiating a shift in approach was challenging, however, given strong opinions within KVM regarding its identity as a 'movement', as Sangeeta describes:

> [M]y objective there was to essentially somehow work a bit on organisational systems, because I felt that was where they were the weakest. They could be more effective, if only they straightened out some processes in terms of better planning, better delivery, you know, being more accountable to what they were doing. And

> it's always a challenge given that, you know, there are these different perceptions of the nature of the organisation [KVM] and in reality also it's all of everything. It's a strange beast. It's a movement. It's an NGO. You need to do some project kind of work, but you also need to allow it to unfold naturally without worrying about deadlines and so on.

Sangeeta was adamant, however, that although taking a 'movement' approach allowed things to 'unfold naturally', it ultimately undermined the accountability of the organisation and its capacity to deliver on outcomes:

> A movement allows you to be accountable to no one else but yourself. Movements are always, I would think, less accountable to anything apart from your conscience. Whereas I think NGO style of functioning at least requires you to be accountable to your *supporters* if nothing else. It should not be that way—the primary accountability should be to the people you have set out to support and help—but at least that force is always at work. You know, deadlines, deliver, do it systematically, plan, review, staffing, all those things …

Sangeeta had made several attempts to introduce a more systematic and accountable approach within KVM but ultimately became frustrated, as improvements did not appear to be taking place. As such, she was ambivalent about whether she would maintain a role within KVM in the future, despite Dutt's renewed commitment to a more professional approach.

In conceptualising a new framework for KVM, a particularly important role was played by two middle-class and highly educated farmers, Ravleen and Ram, both of whom had come to farming from professional backgrounds for a change in lifestyle and to reconnect with issues that they felt were important. Their backgrounds were in business and engineering, respectively, and their analysis played a key role in both framing the nature of KVM's problems and prescribing solutions. A central problem identified by Ravleen and Ram was that KVM's leadership had taken on too many tasks, which had become a barrier to delivering on their promises of making natural farming systems productive. Between 2007 and 2008, KVM had considerably expanded its charter. As awareness spread about the crisis situation in Punjab, Dutt responded by initiating action groups that went beyond the immediate task of facilitating the growth of natural farming. For Ram, this represented a failure with respect to the central goals of the organisation:

> There has been a dilution of the original mandate of promoting natural farming, because a lot of other issues have come up, which has given KVM an opportunity to network with other NGOs in the country. And I think they feel that it's too good an opportunity to pass. So I think the farming aspect has suffered because

> of extra functions being taken on by the same team. And so for the past year, year and a half, I think we've kind of had a standstill or there's even been a slight drop in enrolment or in the spirits of the members of KVM, because we've not been able to get the results, production-wise, that we would have liked … KVM leadership is now more involved in the other avenues. No person is devoted now to actually addressing the problems that have come for the farmers after they have been practicing natural farming.

Ravleen was of a similar opinion, adding that the *ad hoc* manner in which KVM has taken on new roles has been problematic:

> We have expanded our charter very significantly since the organisation was started. It started with just sustainable agriculture. Now you've got health issues, environment, pollution, GM, soil and loads of other things on our plate. And they sort of get picked up *ad hocishly* and put on the backburner as we see fit. There's really no continued momentum that has built up on any one.

At the same time as KVM's leadership was becoming involved in new tasks, a growing number of farmers were experimenting with the techniques they prescribed. It was thus increasingly necessary to maintain contact with a large number of farmers and to ensure they were not disappointed with their results. Because of the new responsibilities KVM had taken on, it was becoming increasingly difficult to attend to the needs of farmers with the kind of dedication that was needed.

During the fieldwork period, several propositions were being discussed regarding how KVM should be restructured to enable goals to be more effectively achieved, especially in relation to farming. A key suggestion was employing more professional staff. In particular, it was suggested KVM should seek professionals who had expertise in ecological agricultural techniques and ideally, had previous experience in assisting farmers in adopting these techniques. It was acknowledged that it may be difficult to find someone with relevant expertise for this task. In Punjab, there are few people with scientific expertise in ecological farming methods as this has not been a priority for the major universities. Ravleen and Ram acknowledged it would be difficult to find someone who would be willing to move to Punjab and to work for what would probably be a low salary.

Both Ravleen and Ram were clear that devoting more time to natural farming should not mean 'dropping the ball' on other issues but, rather, clearly allocating tasks such that work could be carried out more efficiently. In addition to hiring agricultural experts, it was suggested that there should be paid staff devoted to each of the key campaign areas. As Sangeeta noted with reference to limitations she saw in the women's group, WAFE:

> KVM being dominated by men as it is, and Punjab being what it is, [it needs to] put in special efforts to go on patting the place next to you and go on calling the women to come and sit next to you when a group of men are talking. It requires someone to go on saying 'Come, come, come, sit'. But, you know, if you're not even exposed to ideas, some talking, some thinking, how do you expect them to get into action about addressing something? KVM being an all-male group has not helped at all. It means that we need some women activists to be there full time, to be taking that forward. It doesn't happen with part-time efforts, for sure.

By having paid, full-time staff, in contrast to relying on volunteers, input would become more reliable and less *ad hoc*—campaigns could continually apply pressure rather than simply reacting to current events. It would also allow the burden of work to be allocated more evenly and sustainably, helping to alleviate the huge amount of work Umendra and Harmeet had taken on themselves.

Clearly, expanding the number of paid staff would require more money for KVM. The need for more funds was expressed by a large number of interviewees, including those less directly involved in the debates surrounding restructuring. During the research period, there was considerable debate regarding where these additional funds should come from. Other than the money coming from the CGA project, KVM had relied on individual donations. Dutt often collected these himself leading up to key events, such as workshops. It was said by some that the system for the collection of individual donations should be formalised, through membership fees or some system whereby supporters could make regular donations. This suggestion was made only by KVM's urban supporters—it is unclear whether farmers would support the idea of membership fees. Others said that KVM should be more open to receiving funds from other organisations and from donors, though this issue was particularly contentious. It was also said that wherever possible, KVM should be thinking of ways it could produce its own sources of revenue and thus become more independent.

For Ravleen, at least, restructuring KVM should also involve a rethink of factors that had been crucial to its 'movement' identity. For example, reflecting on the amount of his own time that was being consumed in the work of natural farming, he suggests that expecting active participation from farmers in the movement was not sustainable:

> I think the sooner we realise that at least an organic farmer will not have that time to play an active role, the better off we'll be. A conventional farmer still has more time on their hands to be able to do this kind of thing and give time to organisations. But an organic farmer, a sustainable farmer, does not have that kind of time. There's way too much work on the field to be able to disengage for anything more than a short burst. You can ask for short bursts of energy - you

> know, there is a workshop being put together, and you can ask around for people to do just two days of hectic work and they will happily do it. To ask them to do it every day for months and years is too much. And I think that ... we need to get out of that mode and we need to acknowledge that.

Though supportive of, and to some extent driving the restructuring of KVM, Dutt took a cautious approach throughout discussions on how it should take place. He expressed particular concern that putting more resources into workers to demonstrate the yield potential of natural farming would, to some extent, mean playing into the materialistic mentality that he himself opposes. In KVM meetings, there had been discussions of what to do if 'money-centric' farmers come to a more prominent position within KVM. It was feared that such farmers would promote the marketing of organic produce at premium prices and turn the project into a purely commercial endeavour. Dutt and other key members asserted that large land-holding farmers with a commercial mindset should not come to occupy leadership positions within KVM. This would retain the purity of its ideological vision. How this could be achieved, however, was less clear. There has been ongoing struggle to ensure that as new employees come into the organisation, they should not merely be technically competent but also committed to the principles for which KVM stands. Moving forward thus required a careful balance between professionalism and authenticity.

In the time since my fieldwork was completed in early 2010, KVM has made strides towards this more 'professional' approach. They have become far less reluctant about applying for funds from donors and becoming involved in state programmes, but formulate their funding applications in ways that suit their strategic objectives. For example, shortly after fieldwork was completed, they were able to secure funds to employ a full-time staffer to focus on the WAFE project in a sustained way. In addition, they have streamlined systems for collecting individual donations and recruiting volunteers—with an online system on their website that makes it easy for individuals to make contributions. At the 2012 Punjab elections, KVM and other groups placed pressure on political parties to do more about environmental issues. The Shiromani Akali Dal, which won the elections, implemented a series of programmes in cancer mapping, pollution monitoring, and establishing markets for organic food. KVM has involved itself in all of these initiatives, as they have been implemented.

A major turning point occurred in 2012, when KVM featured on Bollywood actor Aamir Khan's TV programme, *Satyamev Jayate*. The programme explores a range of social issues affecting contemporary India, and Dutt and KVM featured in an episode on pesticide contamination. After their appearance on the programme, Satyamev Jayate made a large financial contribution to KVM that allowed them to construct a natural farming centre and head office in Jaito and implement new projects in priority areas. For example, they established regular organic farmers' markets in Chandigarh, Patiala, Bathinda, Gurdaspur, and other major towns in Punjab.[24] This addressed the major concern expressed by farmers experimenting with KVM's techniques that even if they were capable of farming and producing satisfactory yields, they were struggling to reach the market for chemical-free produce. With these funds, KVM was able to provide an arena to respond to this concern.

Having laid down strong foundations as a 'movement' certainly helped KVM establish broad popular support, raise awareness about the issues, and give shape to their own unique moral vision. It positioned them in a sufficiently distant relation to donors and the state for them to develop their own political values and commitments and to win the trust of an NGO-weary public. Coming back to a more 'professional' structure (i.e., to again behave more like an NGO), however, has allowed them to iron out some of the problems with the 'movement' approach—its *ad hoc* nature, its failure to create sustained change in villages, and its urban middle-class bias. With projects to complete in the villages, KVM has become more accountable and has had to focus on delivering sustained outcomes. The funding mechanism may also help to level out some of the effects of KVM's 'rural populist' discourse. Though often having a bias towards neoliberal development models, donor organisations do not want to be perceived to have a bias towards the elite and are likely to steer receiving NGOs towards working with the poor, potentially countering some of KVM's conservative tendencies. It is noteworthy that since 2010, Dutt has more stridently sought to distance himself from his roots in the Hindu Right, perhaps recognising that his nationalist, traditionalist dispositions were not appealing to his potential sponsors.[25]

Conclusion

The case of KVM complicates our understandings of the relationship between NGOs, development, and hegemony. It suggests we must go beyond the simple

---

[24] The fact that these are referred to as 'Organic Farmers' Markets' on KVM's website (http://www.khetivirasatmission.org/organic-farmers-market.html) suggests that reservations about the label 'organic' and the opposition of 'organic' to 'natural' farming may have reduced.

[25] There are, however, counter-claims to this – with some of my long-term informants in the region suggesting that although Dutt may wish to give the appearance of being detached from the Sangh Parivar, since the election of the Modi government in 2014, KVM has been emboldened in its promotion of Hindutva values. These, however, are merely allegations which need to be more systematically investigated.

binary of hegemonic NGOs, who are under the influence of neoliberal donors and the state, versus counter-hegemonic social movements, who represent the authentic perspectives of 'the grassroots'.

Certainly, in the initial stages, KVM's identity as a 'people's movement' allowed it to overcome some of the negative features associated with NGOs. Not being exclusively guided by the objectives of donors allowed them to develop their own (relatively) independent perspective in direct opposition to the hegemonic control of Punjab's food systems by agri-business. It also allowed them to develop direct relationships with a diverse group of constituents and rally broad support for their cause. The formation of 'action groups' can be seen as an expression of this. Action groups allowed KVM to work with various groups concerned with Punjab's crisis scenario, encouraging broad participation and the development of a broad discourse regarding the need for change.

There are, nonetheless, important reasons to be sceptical of KVM's claims to authentic 'people's movement' status, particularly given that 'the people' are not a homogenous entity. The social relations that KVM was embedded within as a 'movement' tended to limit its appeal in rural communities—particularly amongst rural subaltern groups. There were three key reasons for this. First, removing its financial dependency on institutional donors left KVM more dependent on donations from individual members for its financial viability. These donations came mostly from members of the urban middle class, whose perspectives and interests came to exert an exorbitant influence over the organisation. Not only was KVM's focus on rural communities fractured by its involvement in more urban concerns, but its excessive reliance on environmentalist discourses that appealed to an educated urban middle class rendered it somewhat irrelevant in the countryside. Second, partly as a result of this middle-class influence, partly due to Dutt's Hindu nationalist background and partly due to KVM's strategic alliance with land-holding farmers, its discourse as a 'social movement' tended towards agrarian populism. It mobilised romanticised images of the countryside and communal harmony that glossed over class, caste, and gender inequalities. This blocked any potential to use sustainable farming to empower the most marginal. Finally, when KVM did engage in project work, in the form of the CGA, their efforts were undisciplined and diffuse, as activists were too busily involved in other campaign areas. They were ineffective and generated frustration amongst farmers for their lack of sustained support. In brief—while KVM had posed some ideological challenges to the hegemony of agri-business, their position within broader social relations undermined their ability to mount this challenge effectively, while also causing them to subtly reproduce the hegemony of the (predominantly male) middle peasantry at the local level.

The very failure of KVM in trying to implement NGO-style project work while still acting as a movement served as an entry-point for forces wishing to reintegrate it within the mainstream NGO fold. KVM's donors were able to introduce more 'professional' organisational systems and encouraged a restructuring of the organisation. While much of the literature on NGOs might see such disciplinary interventions as examples of the exercise of hegemony in co-opting a radicalised movement (e.g., Kamat, 2002), in some respects they also served to neutralise some of KVM's more problematic relations to traditional hegemonic forces. Not only did they introduce accountability that rendered KVM more effective in engaging with rural communities, but they also tempered the populism of its discourse, which was only serving to empower the traditionally hegemonic middle peasantry of Punjab.

5

The Tamil Nadu Organic Farmers' Movement

The Limits of Participatory Approaches

I was back in green, sunny Tamil Nadu, after a five-year absence, travelling down bumpy roads to a small village outside of the town of Nagapattinam. Agriculture in this village had been badly damaged by the South Asian Tsunami of 2004. I was there to interview members of a grassroots informal collective of around ten to twenty people that had formed in the aftermath of the tsunami at the instigation of Revathi, head of an NGO known as the Tamil Nadu Organic Farmers' Movement (TOFarM). The group formed, in the first instance, to restore tsunami-affected lands but had continued after restoration efforts were complete to provide support mechanisms for people wishing to develop livelihoods through chemical-free, ecological farming. I was accompanied by my translator, Mr. Arockiasamy, and Mathew, who was serving as an unpaid mentor figure within the local collective, providing knowledge and support for others wishing to develop sustainable rural livelihoods.

In the morning, Mathew took us to the large, opulent house of Mudaliyar. As a high-caste, large land-holding farmer, Mudaliyar was part of the traditional elite in a region whose rural society is marked by deep inequalities. Converting a portion of his land to organic cultivation had brought Mudaliyar some benefits. He and many other farmers had been sceptical of the benefits of chemical use for years—yet, until Revathi's intervention, no one had showed them a viable alternative. We sat in the sun on his front porch and heard him speak with great enthusiasm about his success with ecological farming and how the local collective was supporting others as they transition away from chemicals and build sustainable livelihoods. This was what TOFarM was supposed to be about—an initiative, though sparked by an NGO worker, had taken on a life of its own within the community: a self-supporting 'movement' for change.

While we were talking, Lalita arrived. She sat silently with legs crossed at the base of the stairs below the porch and waited. Lalita was an important part of this story, too. She was a Dalit woman, who had mobilised other women from within her caste community to purchase cows, to use their milk for income generation, and sell their dung and urine to organic farmers. It had become a profitable venture for an otherwise asset-less group and was helping to develop the sustainable farming economy.

I had heard about Lalita previously but had not met her. When I realised that it was her sitting so unobtrusively at the base of the stairs, I was excited to greet her. '*Vaanga!*' [Please come] I called out, inviting her with my hand to sit with us. She was unresponsive at first, glaring with a cloudy and uncertain stare. When, after further encouragement from Arockiasamy, she finally came to sit with us, she did so with her body rigid and lips pursed. Yet, there was a steely determination in her eyes and she spoke directly and with confidence about her experiences mobilising local Dalit women.

Mudaliyar excused himself while Lalita told her story. After ten minutes, Mathew arrived with a plate of tea in paper cups and we each took one. Tea is constantly served throughout the day in India, and I thought nothing of the details of Mathew serving us at the time.

After another fifteen minutes, while Lalita was in the middle of telling us about her efforts to represent Dalit women in the local village council, Mathew declared that our interview was over with a crossing of his hands and an abrupt announcement: '*Enough*!' Slightly unsettled by this sudden conclusion to our conversation, we thanked Lalita and said goodbye, while Mathew ushered us away from the house and took us to meet another farmer. We did not see Mudaliyar again.

Mathew's tone shocked me. I had known him to be a very amiable man, and I was surprised at the way he barked the meeting to a conclusion. I kept mulling over it throughout the morning, wondering what precisely had made him become so upset. It was only later in the day, when Arockiasamy and I left the village for lunch that he was able to fill me in on what exactly had happened.

Mudaliyar had excused himself because he felt deeply uncomfortable with a Dalit woman sitting on his front porch. Mathew, in turn, felt so uncomfortable about Mudaliyar's retreat (and the corresponding lack of hospitality) that he ran off to buy tea for us from a nearby store. Hence being served in paper cups—something I thought nothing of at the time. Norms of untouchability meant that had Lalita drank from Mudaliyar's own cups, they would have almost certainly been considered ritually 'polluted'. Suddenly Lalita's steely expression as she gave the interview—concerned, yet bold and resolute—felt all the more poignant.

Having spent the entire morning talking with members of this village about the optimistic prospects of more cooperative and sustainable futures, being informed about the silent transcript of this exchange left me feeling deeply saddened. What hope can there be for an inclusive grassroots movement, when some members cannot even tolerate others to set foot on their property? When they cannot bear the thought of drinking from the same vessel?[26]

———

TOFarM was an initiative to promote sustainable, ecologically-integrated agriculture within Tamil Nadu's rural communities. Though led by an 'urban outsider', the strategic approach of the initiative was to allow rural communities to ultimately direct the movement themselves. The initiative was started by M. Revathi, a middle-school science teacher from the city of Coimbatore, who became interested in promoting ecological farming after becoming aware of the ecological impact of pesticides and issues of farmer indebtedness. Though initially conceived as a loosely organised initiative promoting knowledge exchange, TOFarM rose to prominence after the 2004 South Asian tsunami. Revathi made use of the ecological farming techniques to restore lands damaged by the tsunami in the Cauvery Delta region and has since been active in promoting the continual use of these techniques, so that farmers can reduce their expenditures on chemical inputs.

At first glance, TOFarM would appear to be better positioned to challenge hegemonic relations through sustainable agriculture than an initiative like KVM. For KVM, executive control and moral influence were concentrated amongst members of the urban middle classes. TOFarM, by contrast, though led by an urban middle-class professional, had the explicit aim of transferring control to rural communities so that they could make the 'movement' their own. Furthermore, where KVM showed a bias towards the perspectives of land-holding men, TOFarM made efforts to include subaltern groups, including women, landless labourers, and Dalits. Theoretically, through these measures, TOFarM may have been able to use sustainable farming techniques as a response to agrarian crisis in a way that, rather than reinforcing the privileged position of hegemonic groups, gives voice to the excluded.

TOFarM's approach closely resembles that of the 'participatory development' paradigm. Since the 1990s, the concepts of participatory development (though not always the practice) have become almost mainstream within rural development NGOs, government departments, and transnational institutions (Pretty, 1995). Participatory development emphasises the importance of directly allowing rural communities to partake in and lead the development efforts that are ostensibly for their benefit. This approach to development is positioned as a response to the failures of the centralised, hierarchical, top-down, expert-driven development

[26] Sections from this chapter appeared in Brown, T. 2017. 'Beyond the Participatory Development Impasse: Pathways Forward for Sustainable Rural Development.' *Social Transformations: Journal of the Global South*, 5 (1): 3–30.

system that had proliferated throughout the developing world in the post-colonial period. This kind of top-down approach had been particularly evident in the implementation of the Green Revolution. Chambers (1997), one of the foremost advocates of participatory development, stresses that if development models are to be sustainable, they must include local needs and values. To this end, he suggests that it is crucial to enlist local knowledge and local participation in every phase of development projects. Locals 'plan, act, monitor and evaluate' development projects in alignment with their own needs and aspirations (Chambers, 1997: 102). The normative role of the development worker is also questioned. In the traditional model, the development worker may have presented as a 'leader' and source of knowledge, reinforcing hierarchical relations in project implementation. In a more participatory model, the development worker is more of a 'facilitator' or 'listener', creating a space in which rural communities (particularly those traditionally excluded from development planning) can set their own agenda, take ownership of development projects, thereby making their efforts more likely to be sustained in the long term.

Whether participatory approaches can meaningfully challenge hegemonic power relations has been subject to vigorous academic debate. Scholars have raised at least three key issues. First, some have questioned how much 'facilitators' really can take a 'back seat' in development initiatives, given the imbalances between them and communities in terms of education, class, and privilege (Cooke and Kothari, 2001; Mosse, 2001). Indeed, the fact that they are intervening in villages for the purpose of supposed 'development' encourages rural people to see themselves as 'under-developed', reproducing and amplifying the normativity of being 'developed' as a desirable condition (Green, 2010). From this perspective, participatory development strategies tend to lead to similar outcomes to traditional approaches, while the logic and process of participation obscure the power dynamics at play. Second, it has been argued that participatory approaches typically seek to find local solutions to problems that are embedded in much larger power structures, obscuring the importance of institutional power (Cooke, 2004; Green, 2000). For this reason, participatory interventions have been interpreted as attempts to transfer state responsibilities to communities (Dagnino, 2008). Finally, despite the rhetoric of empowering the marginalised, it has been argued that participatory development initiatives are easily hijacked by local elites, who are better positioned to capture development resources for their own private interests, whilst using the language of the 'common good' (Platteau, 2004).

Through TOFarM, Revathi encountered all the above challenges in adopting participatory approaches—yet she did not give up on their core ethos. Her creative attempts to devise solutions to the problems of participatory development offer valuable insights into whether it is possible for urban outsiders to 'facilitate' a movement for sustainable agriculture that poses a genuine challenge to hegemony.

This is particularly important, for while there are now a multitude of critiques of participatory development, highlighting its incapacity to meaningfully challenge power relations, the literature offers little in the way of alternatives.

Regional Context

Although this chapter explores Revathi's work across a variety of locations, the key focus is on her initiatives in the Cauvery River Delta, on the central coast of Tamil Nadu, in southern India. Ecological agriculture is a radical venture in this region, since, after Punjab, the Cauvery Delta was one of the main targets for Green Revolution interventions in the 1960s and 1970s. Because of its extensive irrigation systems and rich alluvial soils, the Cauvery Delta was one of the regions targeted under the IADP (Frankel, 1971). Tamil Nadu's Green Revolution, however, did not lead to the kind of economic gains that were reaped in Punjab. They also led to more significant social tensions.

Image 5: Farmland in the Cauvery Delta

In the initial phases, there were technical problems. At the centre of Tamil Nadu's Green Revolution were the varieties of rice initially developed at IRRI in the Philippines in the 1960s. As Frankel (1971) documents, farmers were slow to adopt this new variety, as it was seen to be a risk. The variety required a large amount of water for an extended period of time, which the erratic rainfall pattern in Tamil Nadu often did not allow. In 1964, however, local researchers developed a variety that was more appropriate to local conditions. This variety was insect resistant—useful, given Tamil Nadu's pesty tropical conditions—and most importantly, it matured quickly, allowing farmers to obtain a crop during the narrow period in which rainfall was secure, and opening the possibility of double cropping. When the rest of Tamil Nadu was in drought in 1965, this variety, grown in the Cauvery Delta, managed to obtain a sizeable yield, which drew a lot of attention and led to calls that its use be increased throughout the state. Lags in uptake continued, however, as poorer farmers lacked the income to invest in the new seeds and technologies of the Green Revolution. Furthermore, the large number of tenant farmers in the region lacked any incentive for such investments.

More so than in Punjab, the implementation of the Green Revolution in the Cauvery Delta generated significant social tensions that remain relevant to the present day. The region was already mired in high levels of class and caste inequality. Traditionally, rural society in Tamil Nadu has been structured by a relatively rigid caste hierarchy, within which agricultural labour is among the least-valued activities (Alexander, 1975b). Furthermore, as Frankel (1971) outlines, the region had, and continues to have, a high-population density, with large numbers of small-holding farmers and landless labourers. This poor majority were sidelined by the Green Revolution technologies, which provided disproportionate benefits to larger land holders with greater capital to invest. Furthermore, despite increased demand for labour resulting from the more intensive farming practices, an influx of migrant labour from poorer neighbouring districts resulted in a stagnation of local wages. Consequently, landless labourers, who made up a majority of families in the region, received no increased monetary benefit. Thus, in various ways, Green Revolution technologies served to amplify existing social inequality in the Cauvery Delta.

Responding to their relative deprivation, labourers in the region began to assert themselves following the Green Revolution, demanding increased wages and more rights. They were encouraged by the Communist Party of India, which had established a strong presence in some of the coastal towns and villages of the district (Alexander, 1975a). Communist agitations, such as strikes and collective bargaining, led to increased tension between labourers, land owners, and police. The tensions culminated in Kilvenmani village in 1969, where land owners and their supporters massacred over forty people belonging to labouring families (Alexander, 1975a; Mencher, 1974). The fight for some measure of redistributive justice, combined with the backlash from elites, remains a prominent feature of rural politics in the region to this day.

Subsequent developments in the Cauvery Delta have only increased social polarisation and the precariousness of the rural poor. The liberalisation of Tamil Nadu's economy has brought mixed blessings to the predominantly rural delta region. While liberalisation certainly has accelerated rural industrialisation, providing diversified employment opportunities for rural labour (Djurfeldt *et al*, 2008), it has also led to a removal of subsidies for agricultural inputs and deregulation of the rural banking sector (Harriss-White and Janakarajan, 1997). The net effect of these policies has been increased precariousness of agriculture as an economic activity. Additionally, the damming of the Cauvery river upstream has greatly reduced the availability of water for irrigation in the region, further undermining production. As agriculture becomes less viable, small-holding farmers and labourers are increasingly looking to the non-agricultural sector for employment (Harriss, Jeyarajan, and Nagaraj, 2010), leading to high rates of rural out-migration (Djufeldt *et al*, 2008).

In addition to the issues described previously, farming communities in the Cauvery Delta were badly affected by the 2004 South Asian tsunami. On 26 December 2004, a 9.3 magnitude earthquake occurred off the coast of Sumatra, triggering a tsunami that had impacts on coastal communities across South Asia. While Sumatra was the worst affected, there were also huge amounts of damage in other parts of South-East Asia, India, and Sri Lanka. In India, Tamil Nadu was the worst affected state, with 7,923 recorded deaths in the immediate aftermath, 6,023 of which were in the district of Nagapattinam, in the Cauvery Delta (Arya *et al*, 2006). The tsunami created tremendous damage to fishing and agriculture. Whilst substantial funds were mobilised in response to the tsunami by governments, NGOs, and multilateral agencies, there was little coordination between these groups, leading to a somewhat chaotic disaster response (Achuthan, 2009). The tsunami and subsequent salinisation of land in the region led to farmers selling their land at extremely low prices to developers, further accelerating the exodus from agriculture.

Revathi became well known for using ecological techniques to restore lands affected by the tsunami. Yet, she has gone beyond disaster response and attempted to encourage farmers to continue using these techniques to develop sustainable livelihood options. Yet, given the factors described previously, the Cauvery Delta is a challenging terrain for promoting ecological agriculture. Despite a sizeable number of organisations promoting chemical-free farming (Alvares, 2009: 261–297), the shift away from agriculture in the region, particularly post-tsunami, poses logistical challenges. Activists are faced with the double challenge of developing sustainable models and sustaining agriculture *per se*, as vast numbers of people leave the countryside in search of urban employment. In this context, it becomes particularly important to demonstrate that sustainable agricultural methods can provide viable livelihood opportunities for small-holding farmers and landless labourers, alike. This has been one of Revathi's enduring challenges.

Revathi: A Teacher's Journey to Ecological Farming

Before her involvement in activism, Revathi was a middle-school science teacher in Coimbatore, a city in Western Tamil Nadu. Although involved in environmental politics since her student days, her interest in ecological farming began with her participation in a study of bird mortality, which she conducted with the Sálim Ali Centre for Ornithology and Natural History, a research and education centre based in Coimbatore. The project collected the bodies of birds from across India for post-mortem analysis and found high traces of pesticides, such as endosulfan in the birds' heart muscles, which had led to heart attacks and other diseases. Concerned about the potential effects of endosulfan on both bird and human health, Revathi began to explore the reasons for farmers' pesticide use. She did this by interviewing the parents and family members of her students from the nearby countryside.

Revathi's research with farmers was the starting point for her work as a facilitator of ecological farming. She found that farmers were frustrated with the increasing cost of chemical inputs and their resulting indebtedness. Yet, they also felt hopeless and lacked awareness of viable alternatives. Revathi thus began travelling around Tamil Nadu and engaging with people who were practicing chemical-free farming, in the hope that this would provide a model for others. Having seen these farmers' results, Revathi became convinced that farming without chemicals and utilising ecological techniques could provide viable livelihood options. Nonetheless, when she took this information to chemical farmers, they remained sceptical. They were unwilling to take the risk involved in experimenting with new farming techniques, given their already precarious position. Revathi suggested that the farmers should attempt to experiment with ecological techniques on a small section of land and then expand operations if successful, but this suggestion, too, was rebuffed. Most farmers had very small landholdings and, thus, were very constrained in their ability to sacrifice even a small amount of land to conduct an 'experiment'. After several encounters of this kind with farmers, Revathi reached the conclusion that in addition to the challenge of encouraging change amongst farmers too impoverished to take risks, she also had a problem of her own legitimacy. As an urban outsider, farmers were incredulous as to whether she could really understand their position. She concluded that ecological farming would be more authoritative if it were part of a farmers' 'movement', which practicing ecological farmers could lead.

This would be the beginning of an ongoing interest in 'movements' as a method of organising. For Revathi, a 'movement' implied a grassroots association of farmers that would be, in the first instance, a platform for knowledge sharing. By becoming connected with each other and sharing their experiences with different techniques, farmers would be better positioned for success. Through such a movement, chemical farmers who were curious about alternatives could visit chemical-free farms and see the potential of alternative techniques. She also asserted that, by coming together, farmers would have a stronger 'voice' to argue for the effectiveness and importance of their techniques to farming communities, media, and policy-makers. In this sense, Revathi's conception of a 'movement' closely resembles the concept described by Dwyer (2011) as 'grassroots movements': collective platforms that allow poor communities to collaborate to develop their own solutions to pressing social challenges.

Despite hundreds of practicing ecological farmers in Tamil Nadu, Revathi found little evidence of any 'movement' linking them together. After enquiring with farmers, she found that the foremost reason for this was a lack of time. Ecological farming is generally more labour-intensive than its chemical equivalents and, consequently, farmers could not spend time away from their farms to promote the cause collectively. This view was supported by my own interviews with sustainable farmers in Tamil Nadu, some of whom had attempted to promote their perspectives more broadly but found that the demands of farming prevented them from doing this in a sustained manner.

A further reason for the unwillingness to form a movement was the level of divisions amongst those practicing sustainable agriculture in the state. Revathi suggested that even a small difference of opinion on the effectiveness of a particular technique was enough to prevent the most prominent sustainable farming advocates from working together. Interviews with the leaders of other organisations working in Tamil Nadu suggested that these divisions continue, and that they relate to factors other than simply techniques. For example, Pandian, the leader of a small association of chemical-free farmers, made a point of expressing his differences with other prominent leaders in the state on the basis of their farming methods, their methods of organisation, their acceptance of funds, and their level of commitment to the cause. Several other interviews and fieldwork experiences indicated numerous personality clashes within the sustainable agriculture community in the state. These have been an ongoing barrier to the formation of a united movement. Given the sheer number of groups that have proliferated in the state, it is perhaps not surprising that quarrels abound, as groups scramble and compete for funding opportunities.

Revathi was still convinced of the merits of a farmer-led movement, but felt that, by drawing on her skills in writing, teaching, and communication, she might play a role in linking busy ecological farmers with chemical farmers who were interested in alternatives. Thus, in early 2004, she left her job as a school teacher and registered as a society, with the primary objective of distributing materials and demonstrating ecological techniques to farmers. She claims that even at this early stage, she had only intended to play a 'supportive' role in this initiative, as she explained in an interview:

> It should be a farmers' movement. I wouldn't take an active role in that. I mean, everyday work is done by me, but I wouldn't take a position in that, like president, or managing trustee, or any named position. All the roles are taken up by farmers.

As this comment suggests, Revathi was, from the outset, concerned about being perceived to occupy a leadership role. Rather, she hoped to facilitate a movement that farmers would recognise as their own. Before making this long-term committment, however, Revathi felt that she needed to gain a more comprehensive exposure to the agricultural situation in India and an understanding of the strategies being pursued by various movements and organisations. Thus, Revathi and her husband spent the second half of 2004 travelling the country visiting various farmers' organisations to learn more about their methods and the issues with which they were engaged. Some were explicitly promoting sustainable and organic farming, while others were working on farmers' issues and development more broadly. Farmers' movements, such as the Karnataka Rajya Raitha Sagha and the BKU were particular sources of inspiration for Revathi. She saw in them strong movements in which farmers were taking an active role. She was impressed by their capacity to 'raise their voice' and take their message to the streets, forcing state governments to engage with their demands.

At the same time, Revathi identified problems and strategic challenges within farmers' movements. She reached the conclusion that the power of farmers' movements had been stunted by political parties. Various parties had established their own farmers' branches and divided these into subcategories (irrigated farmers, unirrigated farmers, small farmers, large farmers, etc.) as a way of keeping the movement divided and preventing farmers from recognising shared interests and shared experiences of exploitation. This led Revathi to a longstanding concern regarding the divisive role of the state, which would inform her later strategic orientation. Travelling and liaising with diverse farmer organisations around the country provided her with a political education and sharpened her strategic vision. She began to see farmers' issues in a more political light, seeing farmers as a group exploited from all sides, but, due to historical factors, lacking the solidarity required to make collective demands in their own interests.

Although exposed to the diversity of agricultural conditions in India, Revathi also identified certain problems that were common throughout the nation. She claims that wherever she travelled, she encountered indebtedness, pesticide-resistant insects, and a general sense that farmers were 'fed up' with chemicals. These were sources of great distress for farmers. It was at this point that Revathi was able to clarify that the issues she was engaged with were matters of livelihood and food security as well as ecological sustainability. Given the indebtedness and poverty of many farming communities, she recognised that she needed to articulate the socio–economic benefits of ecological farming, rather than simply focusing on the environment:

> R: If you talk simply environment, nothing is going to happen. Environment issues like bird watching or nature walking, they are all having some impact, but

> this problem [of the threat to livelihoods] has a huge potential to operate. This is very important because this is involved with food security and our farmers' lives. And basically environmental balance also.
>
> TB: But you felt environment wasn't enough to mobilise people?
>
> R: No, it won't work out because most of the people were concentrating on economics only, more than environmental ideology. They are very much worried about economics. Whether organic farming will help them to retrieve from their problems. That was their question. They are not concerned about ecology.

Revathi, thus, began to promote ecological farming as an immediate solution to these socio–economic issues, which she claims generated a stronger response. This represented a decisive shift in her approach. Instead of simply promoting her own interest in ecological issues, she would engage with threats to farmers' livelihood, survival, and way of life, in the interest of facilitating a widespread movement for change amongst farmers.

The 2004 South Asian Tsunami: Consolidating the Facilitative Strategy

When the tsunami hit the coast along the Cauvery Delta in December 2004, Revathi immediately travelled there to help in the relief effort. She mobilised her research connections to document salt contamination levels in soil and ground water, which were found to be very high. It quickly became apparent that, while the dominant aid response was directed at fishing communities, the contamination of soil and irrigation waters would precipitate a crisis for farmers as well. Farmers had no previous experience with this level of salinity and, therefore, looked to others in search of answers. Meanwhile, a consensus developed in the research community that it would take between three to five years for the land to become arable again. This created panic amongst farmers, many of whom began to sell their land at desperation prices.

Revathi was confident that, by adopting ecological techniques, it would be possible to reclaim contaminated lands. She developed a chemical-free strategy to remove salinity and aerate compacted soil, which the government allowed her to initiate on an experimental basis in one highly affected village. Her method involved three main techniques. First, salt-affected layers of soil were ploughed and trenches were dug in rows through the fields. This allowed water to pass easily through the soil when it rained, draining into the trench, carrying the salt with it. The salt could then be removed manually from the bottom of the trench. The second technique involved burying hard organic material in the soil, such as logs and sticks. The slow decomposition of these materials helped to rebuild damaged soil. The third technique was to sow a salt-tolerant plant called dhaincha. Dhaincha requires very little water and quickly provided an organic soil cover. This prevented evaporation, which would cause salt to rise back to the top layers of the soil whilst also providing green manure. Revathi claimed that the techniques were successful and, by May of 2005, was able to announce that the lands on which they had been working were restored and ready for cultivation. Following this announcement, local authorities agreed that she could begin work on a larger scale. She began by working on some six hundred acres of land, applying the same techniques.

Working on a larger scale introduced new organisational challenges. This was the first time that Revathi's work had required substantial funds. Money was required to finance machines used in the removal of salt, sand, and mud and to train people in the techniques. Up until this point, she had avoided fund raising and the implementation of NGO-style project work. She had been registered as a trust, adopting a supportive role to the development of grassroots initiatives. Knowing that funds would be required to expand this work, however, she registered as an organisation in March of 2005. Because of the emergency situation, the government had waived the usual requirement that an organisation be registered for three years before receiving foreign contributions. The name chosen was Tamil Nadu Organic Farmers' Movement, reflecting Revathi's continued ideological commitment to 'movements' as a mode of organisation. They successfully applied for funding from a Germany-based donor organisation, which Revathi suggests was supportive of her methods and did not enforce onerous reporting conditions on her.

As the work of Revathi and her team gathered pace, their success began to attract widespread attention from national and international media, academics, agricultural scientists, and politicians. Encouraged by the positive response from outsiders, they decided to extend their work to the entire coast of Nagapattinam, the main coastal district in the Cauvery Delta. To reach larger numbers of farmers, they worked with other NGOs based in the region. They trained NGO workers directly, so that they could take the techniques to communities and followed suggestions from NGOs regarding communities that were likely to be responsive to training programmes. TOFarM has documented that they provided training programmes for at least 16,000 farmers of Nagapattinam District during the latter part of 2005.

To ensure that the impact of these interventions was sustained, Revathi implemented what she termed a 'master farmer' training model. After conducting two-day training programmes with larger groups of farmers, a smaller number were selected to receive more intensive training as 'master farmers'. They were chosen on the basis of their interest in ecological farming, yet, care was also taken that master farmers were not selected from the local elite. These farmers received their training free of cost, on the condition that they take the techniques to others and support them in their transition to ecological farming. Where possible, they would be matched up with a team of others in their village, who were termed their 'fellow farmers', to whom they would provide advice, assistance, and cooperation in developing a sustainable village economy. In Revathi's words, the idea was to 'give [the master farmers] to the community as an asset'. The master farmer would provide grassroots leadership and act as nodes within support networks at the village and district level. They could also maintain communication with Revathi and other supportive outsiders, and call on them for additional help, where required. In the years that followed, this model would become a lynchpin in Revathi's participatory approach, allowing her to provide guidance and support, whilst retaining substantial local autonomy and facilitating the development of leadership-from-below for ecological farming. The 'master farmer' became a trusted local leadership figure, whose results could be observed directly.

In implementing this approach, Revathi kept her efforts as decentralised as possible. This was not only because of her belief in the importance of developing farmers' leadership capacity—it also had strategic value. She had reached the conclusion that the state would be threatened by a movement that would genuinely empower farmers and challenge the power of chemical and seed companies through the promotion of ecological techniques. She saw a decentralised series of micro-movements—small, yet inter-linked collectives of 'master farmers' and their 'fellow farmers'—as an effective way of avoiding unwanted attention from an unsupportive state:

> We cannot say our organisation is a movement. We are now establishing many *movements*. ... Not a single one, a huge one saying ten thousand farmers, ten lakh farmers are there. If we are saying that then it won't be strategically wise ... If you are a big one then, once again, the politicians will take care of you. And they won't allow you to be big and remain, and they will make you collapse. So more small organisations, and more people representing their local level, local locations, it will be better, in these political conditions. That is one of the strategies we are using.

This strategic consideration further reinforced Revathi's light-handed 'facilitative' approach. She would provide support, yet not adopt a centralised leadership role that could potentially make her a target of the state. She was very much aware of the possibility of the state cancelling her right to receive foreign contributions, so did not want to be perceived as a threat, yet still wanted to make a pervasive impact through dispersed, small-scale movements.

By the end of 2005, Revathi felt her team's work in tsunami restoration in Nagapattinam was largely complete. They had attracted a great deal of national and international attention for their work, with visits from former US President and UN Ambassador, Bill Clinton, and former President of India, A. P. J. Abdul Kalam. Clinton went on to recommend Revathi to governments in Sri Lanka and Indonesia, where there was also urgent need for tsunami restoration and both governments extended invitations. Thus, from January 2006 until May of 2009, Revathi and her family spent much of their time travelling around South and South-East Asia, using similar techniques to restore tsunami-affected land. This was followed by another project in West Bengal in 2009 and 2010, in which Revathi assisted farming communities whose lands had been damaged by Cyclone Aila. During this time of intensive travel, Revathi was able to maintain only sporadic contact with the communities she had worked with in the Cauvery Delta. For her, this reaffirmed the need for a light-handed facilitator's approach. She could not provide day-to-day support for farmers—she could only provide training and inspiration and develop their leadership potential so that they could support each other and manage their own 'movement'. As is documented below, the long-term impacts of this approach were mixed.

Facilitating Community-Based Organisations and Regional Federations

Between these various trips abroad, Revathi was employed by an international NGO[27] to consolidate results in the Cauvery Delta and further foster the development of ecological farming through participatory strategies. The NGO imposed certain conditions on how these strategies would be implemented, encouraging Revathi to use a model that they claimed had been demonstrated effective throughout the developing world. The model consisted of forming 'community-based organisations' (CBOs) within villages to work on issues of resource management, with a focus on ecological farming. CBOs were established either for specific tasks, such as preparing bio-inputs or developing livelihood options for landless people, or to empower particular villages or neighbourhoods. These groups would be financed through a 'revolving fund'—essentially a micro-credit programme. Both the NGO and the local community would contribute towards this fund. The CBO would then collectively decide how funds could be distributed as loans to members of the community, specifically in relation to the purchasing of assets for tsunami recovery and transitioning to ecological farming (for example, purchasing cows or building ponds). Once established, CBOs would then be encouraged to 'federate' at a regional level, to form a large governing body, to which the local CBOs would send delegates. After several years of working on this project, Revathi's experience ultimately left her feeling unsatisfied and exposed some of the limitations of participatory development approaches in the context of large international NGO-sponsored initiatives.

The use of community-based organisations in development emerged in the 1990s and has become a key component of the more institutionalised

[27] For confidentiality purposes, this NGO is not named.

participatory development strategies around the world. As Dill (2009) outlines, CBOs are generally unpaid self-help groups, which are intended to function as intermediaries for development interventions. They provide a formal structure through which participation can take place, such that interventions can be more easily documented and groups can be more easily recognised as legitimate by local bodies. Revathi had initially hoped to form CBOs around specific tasks relating to ecological agriculture, such as the preparation of bio-inputs or marketing, but eventually a large number of groups came forward to register as CBOs so that they could take advantage of the micro-credit programme. After facilitating a number of such CBOs throughout the Cauvery Delta region, Revathi encouraged them to 'federate' as a district-level entity. The 'federation' of CBOs has also become an important aspect of participatory development strategies, as it theoretically allows individual CBOs to maintain their autonomy, whilst making the collective decision to pool resources and cooperate with other groups for mutual benefit (Appadurai, 2002). Revathi suggested the name Kadal Oosai for the district-level federation. This name, literally meaning 'the voice of the sea', reflected the coastal identity of farmers and signalled their willingness to 'raise their voice' in the context of regional neglect.

During our interviews in 2010, Revathi made mostly positive comments about her work on this project. She saw it as an extension of her work as a facilitator, which enabled her to provide support and guidance, whilst giving communities the power to ultimately make their own decisions. She suggested that Kadal Oosai had developed as a completely autonomous federation, with each subgroup formulating their own rules, delegating responsibilities, and organising regular meetings: it was, in her words, 'completely managed by the farmers and the landless labourers'. She explained her own, somewhat curtailed role as follows:

> We are not strictly telling any rules and regulations to them. We tell them, wherever there is a need, they have to collectively raise their voice, and collectively purchase things and collectively market their products. For these things we gave our training, our facilitation, and until today we are supporting them.

Despite this early optimism, by the time of our second series of interviews, in 2014, Revathi had become disillusioned with the CBO/federation model. She had three major concerns. The first reflected concerns raised in the wider literature that CBOs tend to form along the lines of caste and kinship groups and hence become exclusionary (de Wit and Berner, 2009; Dill, 2009). In 2010, Revathi had been clear that CBOs needed to be inclusive if *Kadal Oosai* was to function in the manner she had hoped and, to that end, she had explicitly opposed caste, gender, and class discrimination within Kadal Oosai:

> [W]e told them strictly that real involvement of people is our aim, so we cannot support the caste system ... And we told strictly in the elected bodies like the district resource centre, we cannot exclude people from low castes ... They should be given proper space. And women's participation is also very important ... So [we told them] fifty per cent of the space is provided to women.

Despite these attempts to promote inclusivity, by 2014, Revathi was forced to accept that various forms of discrimination had entered into the CBOs, particularly along caste lines. While all CBOs were intended to be open to all castes, it became apparent that over time, they became dominated by particular castes and eventually degenerated into caste associations. In the 'master farmer' model, Revathi could encourage the formation of relations of economic dependency between castes, but when groups were allowed to proliferate on their own, they tended to be more insular. As Revathi explained, 'Whoever could make a group, they joined together—that is, according to their caste, or neighbourhood[28] or according to their status or according to their education'.

When Revathi raised this issue with the international NGO funding the project, they suggested that this could be overcome by having village- and district-level federations, with delegates from all the CBOs theoretically providing representation to all castes within the communities. This, however, neglected the fact that in the context of intense caste discrimination, lower caste delegates would have very little influence or decision-making capacity within the village-level federations. Furthermore, she noted that when community events were held to try to bring the different CBOs together, overt practices of untouchability were observed, with lower caste people being fed from different vessels to the upper caste people. Revathi was highly disappointed by this and felt that it reflected a flaw in the CBO-federation model. As has been noted elsewhere (Dill, 2009), the model rests on the problematic assumption that electoral representation can overcome entrenched discrimination. This neglects the fact that in many developing countries, electoral representation is mired in clientelism and elite bias, which makes it very easy for traditional elites to 'capture' positions of power, both within the localised community-based organisations and in the larger federated bodies.

Revathi's second concern was with the use of micro-credit loans as a method of distributing funds. She strongly believed that money was no substitute for training. For her methods of ecological farming, assets were not as important as knowledge in ensuring success. Furthermore, she had seen that many of the problems affecting rural India had been caused by indebtedness, and that distributing more loans could not be a solution, particularly given that loans were repaid with interest.

---

[28] Neighbourhoods in rural Tamil Nadu are generally segregated according to caste and religion.

She witnessed that many farmers were taking loans from revolving funds for non-productive purposes, such as weddings, which she attributed to a lack of experience in dealing with larger sums of cash. She also heard of communities harassing and abusing people for non-repayment.

Finally, Revathi became convinced that the CBO model was inherently exploitative. The sponsoring NGOs had demanded rigorous book-keeping from the CBOs, which most rural people lacked the education and skills to maintain. They were therefore forced to employ outsiders to do their book-keeping, using money coming from their collective fund. This left Revathi feeling particularly uneasy. Furthermore, whether giving up their own time, or employing an outsider, Revathi began to question whether these book-keeping tasks constituted a service that the rural people should have been able to expect from the state or banks. She questioned who the real beneficiaries of such operations were and suspected that the funding NGO may have come under the influence of vested interests. The small amount of money disbursed through micro-loans was not sufficient to make significant improvements in people's lives. It would, however, make a significant contribution to financial institutions, as collective funds served to consolidate dispersed rural savings, without the substantial associated book-keeping costs (these costs being borne by the communities themselves). When Revathi raised this concern with the sponsoring NGO, she was told that her attitude was 'immature' and that this model had been tested and was being used all over the world. Ultimately, however, Revathi's view was vindicated. Farmers came to reject the CBO model. In most cases, they divided the group savings and distributed them amongst members, after numerous problems with repayments.

The sponsoring NGO's continued insistence on the CBO model, despite clear and reasonable negative feedback from Revathi clearly shows the problems associated with excessive donor influence. Donors seemed incapable of adjusting their strict adherence to particular models, despite clear signs of failure with respect to stated objectives. This highlights the limitations of donor-dependent NGOs in facilitating change that effectively challenges hegemonic power relations. According to Revathi, donors were unresponsive to her concerns that upper caste power was being reinforced or that the CBO model was exploitative, which raises questions regarding their underlying agenda.

After this experience, Revathi became more convinced of the merits of her 'master farmer' model. This model was not based on representation, which could be subverted by the powerful, but on economic empowerment and cooperation. Having a master farmer and set of fellow farmers from various local caste groups encouraged relations of mutual dependence and support within the community. Furthermore, it provided communities with what Revathi believed was the most crucial factor required to transition to ecological farming: technical knowledge.

A well-trained, well-equipped master farmer could fill in gaps in project implementation, identifying problems at the grassroots level and requesting outside support where required. She firmly believed that with the CBO model, much of the technical know-how and economic empowerment was lost, as at least half of communities' energy was spent in the formation and maintenance of the groups. The facilitation of political empowerment was surely a priority, yet Revathi believed that this would evolve naturally, as communities became more empowered at the grassroots level, by cooperatively taking control of their farming systems.

The case of Kadal Oosai highlights the very different ways in which 'facilitation' and 'participation' can be interpreted in the context of participatory rural development projects. Revathi conceptualised her primary task as facilitating 'movements'. By this, she implied building unity amongst farmers, their capacity to share knowledge, and 'raise their voice' if their rights were infringed. For her donor, however, the objective was to facilitate the formation of CBOs and microfinance initiatives, developing communities' capacity to provide services for themselves. This discrepancy lays bare the fundamental ambiguity in the facilitator's role within the participatory development establishment and the fact that *what* one facilitates is never simply the choice of the 'empowered community'. The facilitator's input always has some impact on the choices that are available to rural communities. Given this, understanding the outcomes of facilitative approaches to rural development requires careful attention to the institutional context in which development organisations are embedded and the interests being served (Brown, 2017).

The Demonstration Farm

As part of her efforts in the Cauvery Delta, Revathi wanted to offer practical demonstrations of the potential of a diverse range of ecological techniques for income generation. Thus, in September 2006, Revathi and her family purchased twelve acres of low-lying land that was still badly affected by the tsunami, with dense soil and stagnating saline water. They set about establishing it as a demonstration farm, highlighting the diverse high-value crops that could be grown in a region in which most farmers had been growing only rice since the time of the Green Revolution.

The farm, which they named Annam Amudham,[29] was not a commercial enterprise. Its main output was organic seed, which was given freely to farmers on the condition that they would, in turn, give twice as much seed to other farmers

[29] In Tamil, *Annam* means simply 'food'. *Amudham* is the Tamil equivalent of the Sanskrit word *Amrit*—a mythical substance which, if consumed, gives eternal life.

after their harvest. The farm generated some income through the sale of crops at local markets, and this was all spent on wages for farm workers and in further developing the farm and its facilities. A small number of agricultural labourers were employed and provided with housing on the farm, as well as a master farmer, Murugan, whose role was to manage the farm while Revathi and her family were elsewhere. Revathi prioritised providing fair conditions for the labourers on the farm. Revathi paid low-caste farm labourers considerably higher wages than the local rate, perceiving the local rate to be exploitative. Furthermore, she paid female workers as much as male workers, which went against local custom. Indeed, Revathi preferred not to use the term 'labourers' for those who worked on her farm, as it came with an attached social stigma. Instead, she used the term 'co-producers'. Surplus food was consumed by those working on the farm. In an interview, one labourer working at *Annam Amudham* listed as a key advantage to working on there the fact that, in addition to her wage, she received food, including fruits and vegetables, for her labour, which improved her family's diet. She also owned two acres of land herself and was able to learn a range of ecological techniques through TOFarM to apply on her own land.

Image 6: TOFarM's 'Annam Amudham' demonstration farm

With time, however, the farm began to cause considerable tension within the local community. Revathi claims that she received complaints from local farmers who relied on wage labourers. They felt that the conditions that TOFarM were providing for labourers, women in particular, were setting a bad example and would raise the expectations of others. Furthermore, despite providing good wages, Revathi and her family's unfamiliarity with local conditions meant that the relations with workers were often strained and uncertain. For example, Revathi felt unsure what labourers were doing with the food surplus when they were away from the farm site (which was a considerable amount of the time) and suspected that it was being sold without their knowledge on local markets. Furthermore, TOFarM's insistence on not discriminating on the basis of religion created additional tensions. Murugan, the farm manager, was a Muslim, while the labourers were Hindus and resented taking orders from him.

These communal tensions expressed themselves during fieldwork in early 2010. A workshop had been organised, aimed at farmers from around the district, but particularly those from the nearby village. The farm labourers were asked to go to the village and share the invitation with all farming families. When they went to the village, however, the workers only spread the invitation within their own neighbourhood. Since villages are spatially divided along lines of caste and religion, this effectively meant that the workers only invited Hindus and people of their own caste. The potential for communal tension was strongly felt. Although Revathi and her family were not religious, they were of Hindu background and feared that the selective distribution of invitations would be perceived as an affront to the Muslim community. The police were notified of the potential for communal conflict, and the house was kept under security for the night. Although no conflict ensued, the event was highly stressful for Revathi and, at the time, she spoke about abandoning the farm in the Cauvery Delta and establishing a farm closer to her home city of Coimbatore, where she believed caste and religious issues would not pose such an obstacle to her work. She suggested that these kinds of communal considerations always needed to be taken into account at the Cauvery Delta. The incident made clear some of the challenges of urban outsiders attempting to facilitate a socially just and inclusive movement within highly divided rural societies. Without in-depth familiarity with the local social environment, navigating the complex terrain of social power poses a bewildering array of challenges.

Long-term Impact in the Cauvery Delta

During fieldwork in late 2014, I visited a village outside the town of Nagapattinam in the Cauvery Delta which had been a target of Revathi's restoration efforts. In-depth interviews with seven villagers not only allowed Revathi's claims to be cross-checked but also provided great insights into which aspects of her work had enduring impacts. The findings provided support for Revathi's claim that a focus on economic empowerment would have a greater impact than micro-credit programmes, CBOs or regional federations, though the reasons for this may have differed from those that Revathi had supposed.

Revathi's 'master farmer' model had clearly had an impact in this village. Mathew, a retired public servant with five acres of land, had been selected as a master farmer by Revathi shortly after tsunami recovery work began. He was chosen due to his interest in ecological farming and leadership qualities. With Revathi's facilitation, Mathew established a network of some ten to twenty 'fellow farmers' from his village, whom he supported in their transition to ecological farming. He also subsequently become involved in a range of campaigns in defence of farmers in the region, and at the time of fieldwork in 2014, was deeply involved in a campaign to increase water allocations to the Cauvery Delta. At the time of research, Mathew's network of ecological farmers continued to maintain contact and supportive relations, though were not meeting as frequently as in the past. The members of this group who were interviewed, including Mathew himself, referred to Mathew as the 'guru' of the group, who provided advice whenever needed to his 'disciples'.

As the anecdote that opened this chapter demonstrated, there were certainly barriers to inclusive participation within this support network, yet, importantly, it did not appear to be dominated by any particular community. Revathi had facilitated this collaborative group in such a way that it was composed of members of various castes and religious groups and both small and large land-holding farmers. One particularly prominent member of this network, a Dalit woman named Lalita, developed as a leader in her own right. Recognising the need for more cows in the village to provide bio-inputs to support ecological farming, Revathi had encouraged landless households to revive their traditions of animal husbandry. Animal husbandry had become a stigmatised livelihood in the region, due to associations with illiteracy and low-caste status. With Revathi's backing, Lalita encouraged landless people in the village to recognise the dignity in this form of labour and promoted keeping cows as a means to generate livelihoods. With some financial support from the 'revolving fund', these communities had purchased cows and were making income through the sale of milk to consumers and cow dung and urine as inputs for ecological farmers. The relations of mutual dependence that this created helped to sustain connections and positive relations within the group. Like Mathew, Lalita had also developed as a leader on other issues, representing Dalit women at the local Gram Sabha.[30] This highlights the empowering effects of Revathi's interventions for members of various social strata.

In terms of the technical practices of ecological farming, the farmers who received training acknowledged that Revathi's input was more than adequate. One small-holding farmer, Madhu, was particularly vociferous on this point, advising that he had already gained much from Revathi. Indeed, despite a lack of

[30] Gram Sabhas (or Gram Panchayats) are local elective governments in India's villages.

formal education, he was able to recite the technical knowledge of agro-ecological processes that Revathi had imparted, years after her departure. Emphasising that he did not expect or require any further support from Revathi, Madhu stated that

> Training is over. They have given us everything we need—they taught us how to cultivate. I'd say that that is what we are doing now [cultivating]. After studies, you go out to work.

The continuation of this network of farmers and their confidence in practicing ecological farming provides evidence in support of Revathi's interventions; yet, comments from villagers also suggested some limitations to her approach and of her participatory methods. Perhaps, Revathi's most problematic assumption was that farmers would be able to take over the leadership of the collective structures whose formation she had facilitated. Villagers noted that Revathi's involvement with their communities was intermittent and decreased over time. She helped to facilitate group formation between her time doing tsunami and hurricane restoration abroad and elsewhere in India. From 2008 onwards, her role was highly truncated. Firmly holding to the view that her role was to provide rural communities with the tools required to organise and lead their own 'movements', it was important to her that her role would become less significant over time. Nonetheless, this was at odds with community expectations. Villagers spoke highly of Revathi's work in tsunami restoration, yet, perhaps because of this positive impact, there was a sense of abandonment when she withdrew. This was clearly articulated by Mathew, who attributed the collapse of village-level collectives and Kadal Oosai to the fact that Revathi left so abruptly. He reported that after Revathi's departure, rather than farmers taking over the leadership of Kadal Oosai, a new NGO named BRIGHT[31] arrived on the scene. This new NGO was unable to provide adequate support.

> While Revathi was here, everything was smooth. Then it changed. After BRIGHT came, everything was broken. While she was there, she took care of us, but after she left, nobody bothered. BRIGHT was there, but Revathi, when she left, she washed her hands of us. Not even a phone call in eight years.

Given that, earlier in the interview, Mathew had referred to Revathi as *Amma* (mother), the use of the phrase 'washed her hands' (*kai kazhuvirathu*) is particularly poignant here, implying a relinquishing of maternal responsibilities. The sense of abandonment was palpable. In his view, more work needed to be done, so that he could take over the coordination process and promote ecological farming more

[31] A pseudonym

effectively.[32] He felt that BRIGHT was completely unable to replace Revathi as a facilitator of Kadal Oosai, emphasising that its leader lacked her mental maturity (*manapakavaum*) and financial backing. It was suggested that BRIGHT had used Kadal Oosai as a means of generating income for itself, with NGO workers charging communities book-keeping fees for managing their collectives. Farmers were aware that this was exploitative and unimpressed by BRIGHT's leadership, which led to the eventual collapse of the Kadal Oosai federation.

The collapse of Kadal Oosai suggests another reason for the weakness of Revathi's model in the current institutional climate. On the one hand, communities were not entirely equipped to take control of this movement for themselves, refuting a core assumption of the participatory strategy—namely, that leadership can eventually be completely transferred to communities. Indeed, communities evidently expected Revathi to stay on and provide ongoing leadership and guidance. On the other hand, there were other NGOs who stood to profit from taking the collective structures under their wing, continuing relations of dependency and patronage. Once again, institutional context cannot be ignored. Other NGOs will always be present and may intervene in ways that undermine processes of grassroots empowerment.

While the 'Master Farmer' model appears to have been more successful in the long term, there were some signs that this, too, could have benefitted from more ongoing support from Revathi. It was certainly an achievement that in 2014, ten years after the tsunami, the collective of ten to twenty farmers in Mathew's village were still maintaining their supportive relationship and continued practicing ecological farming. Yet, not all farmers in the village were doing so. Indeed, it seems that beyond this ten to twenty persons, most in the village had gradually returned to chemical farming. Mathew claimed that this occurred at the same time as Revathi was withdrawing. This created problems even for practicing natural farmers, as the use of flood irrigation in the paddy-growing region had led to the cross-contamination of their fields by their neighbour's chemicals, making organic certification impossible.

An even clearer indication of the need for more sustained contact came from Lalita. Lalita's determined efforts to encourage local women to purchase cows for livelihood generation were ultimately undermined by external circumstances. She had convinced more than ten local women to save money to purchase cows and she, herself, was raising four. Yet, this was to be tragically interrupted when an outbreak of foot and mouth disease resulted in the deaths of all of Lalita's

[32] This resonates well with Ricks' (2016) recent study, which found that regular interaction was more likely to lead to positive outcomes in participatory development projects than training programmes.

cows and of many others in the village. The others who had bought cows at her suggestion felt betrayed, and this, she said, had undermined the collectives that were developing within the village. Some ongoing support from Revathi may have assuaged the economic and social backlash that Lalita encountered as a result of these circumstances. This demonstrates that participatory strategies do not inevitably lead to outcomes that are empowering. It cannot be assumed that after a brief intervention by development workers, communities can be left to take over, particularly given the unpredictability of the natural and social environment in which they operate.

The caste discrimination against Lalita observed during fieldwork also provides some evidence of the need for sustainable agriculture activists to have a more long-term engagement with communities, if they are to challenge the entrenched power of rural elites at the local level. As was mentioned in the introduction to this chapter, although Mathew and his 'fellow farmers' were engaged in relations of economic cooperation with Lalita, some members of the collective were clearly continuing to observe traditional rules of untouchability with regard to her and her husband—by refusing to serve her tea from their own cups, for example. This suggests limits to this effort's empowering effects: economic or even political cooperation will always be limited when extreme practices of social exclusion persist. One wonders whether Revathi might have played some role in continuing to challenge such practices, which rural communities themselves may have come to (perhaps reluctantly) accept.

Conclusion

Revathi was aware of the problems associated with herself, as an urban middle-class outsider, being perceived to occupy a leadership position within TOFarM. There were three key reasons that she felt the need to develop an organisational model within which her role was relegated to that of a 'facilitator', rather than a 'leader'. First, she believed that if farmers could build a movement for ecological agriculture that responded to *their* needs and that they could recognise as their own, then its impact would be far wider and more sustainable than anything that she could direct. Second, she saw that, as an outsider, her understanding of the harsh realities faced by rural communities could easily be questioned; she needed local, rural leadership to demonstrate that her suggestions could bear fruit. People needed first-hand demonstrations that poor farmers could achieve positive results with ecological farming. Third, in a context in which Revathi believed the state was not supportive of her agenda for farmer empowerment, a centralised leadership model could make her a target, which, if successfully attacked, could quickly cause the entire effort to collapse. It was better to develop more decentralised (yet interconnected) grassroots leadership at multiple nodes throughout the region.

Revathi's approach, which had close affinities with the ethos and methods of the 'participatory development' paradigm, allowed rural communities to take more of a directive role in articulating a counter-hegemonic response to agrarian crisis than had been the case with KVM. Furthermore, by facilitating the participation of subaltern groups in the process (women, dalits, and landless labourers), Revathi made efforts to ensure that this response did not merely reproduce the hegemonic position of the rural elite.

When donor funds became involved, however, opportunities arose for an ostensibly rural-led, participatory initiative to be manipulated to serve hegemonic interests. Tensions between Revathi and her sponsoring NGO over the effectiveness of CBOs and micro-credit programmes aptly illustrate this. Revathi became convinced that the CBOs were ineffective and reproduced the domination of upper castes and that micro-credit programmes were inappropriate, given that farmers were already struggling with debt and loans were not of a sufficient size to lead to profitable returns. For Revathi, her sponsor's continued insistence on these approaches, despite evidence that they were not contributing towards stated aims, revealed ulterior motives. The ultimate goal seemed to be to reduce the costs of administration and to help consolidate rural savings for the finance sector, rather than to meaningfully empower the rural poor.

These struggles with her sponsors ultimately led Revathi to consolidate her own belief in the value of her master farmer model. A model that focussed on building relations of mutual economic interdependence and empowerment seemed less vulnerable to political co-option by elites than the CBO model. Certainly, the endurance of the collective interviewed at Nagapattinam, years after Revathi's departure, suggests that such interventions can lead to meaningful change as they build livelihoods founded in cooperation that traverse traditional community divisions. Commendable as this model may be, however, this case study suggests various ways in which any effort at facilitation will face challenges when attempting to intervene in deeply divided rural communities. Revathi, as an outsider unaccustomed to the rural social landscape, struggled to navigate the often sensitive terrain of inter-communal relations. It may have also been unrealistic to expect genuine ties of solidarity to develop in communities in which practices of radical social exclusion prevail. I draw on Revathi's example to consider the possibility of participatory models of development that are genuinely empowering for the rural subaltern in greater detail in Chapter 7.

6

The Beej Bachao Andolan

How 'Grassroots' is the Grassroots?

I was sitting in a sweet shop in the market town of Khadi in early 2014, eating jalebi, sipping on a smoky, over-sweetened cup of tea. I was on my way to visit the village of Rampur, just outside of the Henwal Valley, which had been my base in 2010 while I conducted my research on the Beej Bachao Andolan (Save Seeds Movement, hereafter BBA). I was returning to the village to spend some time with the family with whom I had stayed and to see how things had progressed for the movement.

Khadi was the closest town to the main road. From there, one had to board a 'shared taxi'—a ten-seater jeep that would be filled with as many as eighteen people (and often a few more on the roof). The taxi would meander up dirt roads to the villages further up the hills. My friend and translator Deepak had been told that it would take a while for the jeep to fill and that we should just relax for some time.

As I sat in the sweet shop, Deepak ushered in a familiar face from outside. It was Dhoom Singh Negi, who I had interviewed for my research on BBA during my previous visit. I was very pleased to see him. Dhoom Singh has a gentle manner and warmth of presence that immediately sets one at ease. He was a prominent activist within BBA but perhaps more well-known for his work as a major organiser of the Chipko movement of the 1970s and the anti-Tehri dam movement of the 1980s and 1990s. He is known locally for his work as a Gandhian social worker and widely respected as an honest and humble man who has sincerely put Gandhi's philosophies of simple living and locally-oriented social work into practice.

Deepak reminded Dhoom Singh of who I was. We exchanged pleasantries and he asked me about my work. I told him that I had recently had an article published about my research on BBA (Brown, 2014). I told him what I had written: that BBA had, in many ways, kept the Gandhian and ecological ethic of the Chipko movement alive in the Henwal Valley—but that this ethic did not seem to resonate with the younger generation, many of whom were leaving their villages due to a severe shortage of employment opportunities.

Dhoom Singh was courteous and considered in his reply. He acknowledged that young people in the region were not interested in Gandhian development. Most of them are moving out in search of greener pastures. The problem, he said, is that they are all following a different model of development, in which cash income is the measure of progress. However, he insisted that sooner or later, people would need to return to Gandhi. The current development model, he said, is dependent on the rapid exploitation of natural resources. Since this is occurring at a rate much faster than the rate of natural regeneration, it cannot last forever—eventually it will need to be re-thought. People would return to agriculture.

Dhoom Singh finished his tea and had to bid us farewell. Soon after, Deepak and I squeezed into the jeep with fourteen other people and made the uncomfortable journey to Rampur. As we winded our way along the perilously narrow road, I wondered about what Dhoom Singh had said. His position was difficult to argue against. There was no way that the capitalist path of growth could continue indefinitely on a finite planet. It was not, strictly speaking, sustainable. Yet, despite the fact that what Dhoom Singh had said was clearly *true*, this was the kind of truth that—while everyone would accept it intellectually—very few would ever *take seriously*.

In a world founded on *maya*, irrelevance is the price you pay for strict adherence to the truth. This was BBA's enduring challenge.[33]

BBA is an initiative led by a small group of small-holding farmers in the Garhwal Himalaya, in the state of Uttarakhand. It was initiated in 1985 by Vijay Jardhari of village Jardhargaon, in District Tehri-Garhwal. Jardhari had been an activist in the Chipko movement, which mobilised against commercial forestry in the region in the 1970s. BBA was, in part, an application of Chipko's Gandhian and ecological ethos to local agricultural development (Brown, 2014a). BBA's primary focus is the preservation of traditional varieties of seed and convincing local farmers to reject hybrid-variety seeds. Its members stress that the traditional varieties of seeds of the region produce crops that are hardy, extremely rich in biodiversity and well adapted to local conditions. Concerned that hybrid seeds could be the entry point for chemical agriculture more broadly, they see this struggle as being crucial to preserving the traditional agriculture and way of life of the hills. To this end, BBA are active within their own villages and local institutions convincing farmers to continue their traditional system of farming.

BBA has been successful in reclaiming a large number of traditional seeds, whose survival was threatened by the Green Revolution in the 1960s and 1970s. It has developed a seed bank with hundreds of varieties of local seeds and ensures these continue to be available to farmers. It has also developed many links with other organisations with similar focus regionally, nationally, and globally and

[33] Sections of this chapter appeared in Brown, T. 2014. 'Chipko legacies: sustaining an ecological ethic in the context of agrarian change.' *Asian Studies Review*, 38 (4), 639–357.

participates in the articulation of sustainable agriculture discourses at all three levels. It has done this while working in a largely informal manner and with very meagre access to funds.

BBA differs to KVM and TOFarM, in that its leaders are *not* urban middle-class 'outsiders'. Its leaders were all born and raised in the villages within which they work and derive their livelihoods from farming. This being the case, one might expect that they are better positioned to understand local issues, less under the influence of external hegemonic forces and, therefore, better positioned to use sustainable farming systems in a way that reflects local needs and values. Nonetheless, as I show in the chapter, BBA has struggled to sustain relevance in the contemporary context. I show this is partly attributable to the fact that their message is now widely recognised as common sense but also because the region now faces new challenges, such as climate change and accelerated rural out-migration. I conclude by arguing that BBA's failure to meaningfully engage with these new issues is partly attributable to the links that they have developed with national and transnational activist networks. BBA has found more fertile opportunities to spread their message through these networks, largely dominated by global middle-class activists, than in their own villages. This may have led to a partial neglect of more pressing local issues, precluding BBA's potential to continually express a counter-hegemonic message from the grassroots.

Regional Context

BBA operates predominantly from a group of villages located in and around the Henwal Valley, in District Tehri-Garhwal, Uttarakhand. Some of these villages are close to the river valley and have access to irrigation facilities; others are further upland and their farming is entirely rain fed.

The state of Uttarakhand was formed from the mountainous section of North-Western Uttar Pradesh in 1999. The state is divided into two regions, Kumaon in the East and Garhwal in the West. The two regions are divided in terms of language, culture, and colonial history. Kumaon was administered directly by the British during the colonial period, whereas Garhwal was ruled by the Raja of Tehri. BBA is located in the Garhwal region. As will be elucidated in the next section, the political history of the region has a strong bearing on BBA's activism.

Agriculture in both regions is a precarious activity, easily disturbed by changes in the local environment. Due to fragile soils, agriculture is highly dependent on forests. Forests are a source of fodder and green manure and help to retain soil moisture. People have, consequently, developed sophisticated practices of forest management to support their agricultural practices (Farooquee and Maikhuri, 2007). Farming is mostly oriented towards the subsistence needs of local communities, rather than the market. Steep slopes provide a barrier to extensive cultivation, and fragile soils limit the potential for intensive cultivation, constraining the development of commercial agriculture (Guha, 2000: 27–28). Indeed, only fifteen per cent of farmers in Uttarakhand grow commercial crops, and this figure is far lower in the more mountainous districts (Pande, 1996). BBA, therefore, operates in a very different context to KVM and TOFarM, dealing to a far greater extent with subsistence-oriented farmers who have close connections to local forests.

Image 7: The Henwal Valley

Uttarakhand was never a major target for development under the Green Revolution. While there were some attempts to bring the High Yielding Varieties Programme to the irrigated valleys of Uttarakhand, there was never much potential for their application in unirrigated upland areas (Krishna, 2002). Farmers interviewed for this research stated that during the 1970s, urea became available in shops in small towns and that these continue to be used occasionally by some farmers. One elderly interviewee for this research suggested that a major role was played by cooperative societies, who gave loans to farmers on the condition that they take and use urea on their fields. Despite these attempts, evidence suggests that the use of hybrid seeds, chemical fertilisers, and pesticides remains low in Uttarakhand (Krishna, 2002). In recognition of the fact that many farmers in the state are thus 'organic by default', in 2001, the state government declared

Uttarakhand an 'organic state', and established initiatives to have its produce recognised, certified, and marketed as organic (Organic Uttarakhand, 2012). This enabled certified farmers to sell their produce on both domestic and international organics markets, offering a new source of upward mobility for some of the region's farmers (Galvin, 2014).

Agriculture in Uttarakhand has been destabilised by a range of contemporary developments. In the course of the last decade, successive governments have pursued a development agenda based on harnessing the hydro power of the hills, constructing a huge number of dams of various sizes and exporting electricity to cities on the plains. The construction of these dams has displaced thousands of farmers, who have limited access to alternative livelihoods and less than a quarter of whom were adequately resettled (Agrawal, 2013). Climate change is also a major issue. Agriculture throughout the Himalayan region is particularly vulnerable to climate change, as glacial melts and more erratic monsoons make water access a major concern. The construction of roads and dams has also made Uttarakhand vulnerable to severe erosion and dangerous landslides in the event of flooding, as was seen during the devastating floods of 2013 (Balasubramanian and Kumar, 2014).

Uttarakhand has a relatively egalitarian social structure in comparison to much of the rest of India. Guha (2000) argues that this is a consequence of the ecological constraints on agriculture, described previously. Since the food produced in the region is mostly consumed by cultivators and their families, there has been little potential for the accumulation of surpluses. In the absence of surpluses, there has been little opportunity for any group to acquire large tracts of land. As such, the class structure of the region is somewhat flat. Most farms are cultivated by those who own the land; tenants and agricultural labourers are very few in number. Furthermore, the difficult terrain constrains families' capacity to provide for themselves based on any one economic activity. The family has had to act as a versatile economic unit, with all members taking on a variety of roles. Consequently, women have taken a more active role in public life, most visibly through their participation in agriculture. Additionally, caste is not easily defined by professional markers, as members of all castes participate in most forms of economic activity. Guha (2000) suggests that this may explain the lack of caste prohibitions and rules of untouchability in the Himalayas compared to the plains.

The lack of a decisively 'elite' class in rural Uttarakhand safeguards against the threat of 'elite capture' of sustainable farming projects discussed in previous chapters. The relatively flat social structure is another reason why BBA is less at risk of being subsumed by locally hegemonic groups. Yet, Uttarakhand's surface-level egalitarianism cannot be taken for granted. Women's participation in the workforce, for example, though giving women greater public visibility than on the plains of India, does not translate into positive living conditions. Indeed, in a survey of women's labour in the more mountainous districts of the region, Pande (1996) finds that women are not only responsible for most agricultural tasks but have exclusive responsibility for domestic tasks. On average, women in the region work between sixteen to seventeen hours per day, and the drudgery of their tasks takes a considerable toll on them. Pande's findings also suggest that despite their greater participation in the village economy, hill women have very limited roles in household decision-making and local politics. On matters of caste, although overt practices of caste discrimination may not be prominent, it is worth noting that a major catalyst for the movement for an independent state of Uttarakhand was an agitation led by upper castes opposing reserved seats for Dalits in the Uttar Pradesh parliament (Kumar, 2011). This clearly suggests that caste is far from an insignificant factor in Uttarakhand. These factors should be kept in mind in discussing the case of BBA, which, while not engaging in activities that support or reproduce gender or caste hierarchies, has also done little in recent years to directly challenge them.

History of Struggle in Garhwal

The Garhwal region has a rich history of social movements that dates at least as far back as the colonial period. BBA is situated within that history and consciously draws on it. Garhwal's long tradition of protest is well illustrated by the practice of *dhandak*. Under the rule of the Raja of Tehri, the peasantry took to *dhandak* when they felt that local officials were acting injudiciously. This involved refusal to cooperate with officials and, in some cases, would end with peasants rallying to the city of Tehri to have their grievances heard by the Raja.

Both Kumaon and Garhwal were major sites in struggle against British colonialism (Handa, 2002). In Garhwal, the Independence struggle introduced Gandhian ideologies, particularly through the prominent Gandhian activists Sri Dev Suman and the young Sunderlal Bahuguna, who were drawn to Gandhi's ideas related to localised village-level development. After India's Independence, Bahuguna established a Sarvodaya[34] collective in Garhwal, which facilitated

[34] The term 'Sarvodaya' was coined by Gandhi and first used as the title for his abridged translation of Ruskin's *Unto This Last*. Gandhi renders the term back into English as 'the welfare of all', though more literally, it could be translated as 'universal upliftment'. The concept speaks to the value Gandhi placed on social work grounded in the ethics of independence, self-reliance, and village level development aimed at the moral and economic upliftment of all people. Gandhi stressed that this goal would not be achieved through development of theory, and encouraged his followers to discover the truth of this philosophy through 'constructive work' in communities. As Kantowsky (1980) outlines, the Sarvodaya

cooperative development at the village level and opposed practices of untouchability and caste discrimination (James, 2013). In Kumaon, Gandhian socialist ideas influenced by Vinoba Bhave held greater sway, and this was reflected in the establishment of the Lakshmi Ashram by Sarala Devi in 1946, which explicitly focussed on the economic upliftment of women and creating a society based on principles of equality (Klenk, 2010).

This collective, along with several others that formed in Uttarakhand, fed into the Chipko movement, which developed in the region in the 1970s. The Chipko movement is highly important to BBA, as its leading activists began their activist careers in Chipko and continue to be informed by its ethos (Brown, 2014). Guha (2000) situates Chipko within the context of intensifying commercial forestry in the post-Independence period. With growing demand for timber as industrialisation gained pace, the hills increasingly became a source of raw materials. Significantly, very little of the wealth generated through forestry went back into hill communities. Forest workers were usually non-local and processing occurred in factories on the plains. The loss of forest cover was also causing environmental damage in the form of severe erosion, which put agriculture and human life at risk in the event of floods (Weber, 1988).

The Chipko movement began in earnest in 1973. Chipko is often referred to as an exemplar of the 'environmentalism of the poor' (Guha and Martinez-Alier, 1997). Its activists are said to be the 'original tree-huggers', who, according to mainstream narrative, embraced the trees to protect them from loggers. It is also occasionally positioned as an 'eco-feminist' movement. Shiva (1989: 68), for example, describes Chipko as being 'a resurgence of woman power and ecological concern', derived from an essentially female understanding of the holistic value of forests and the importance of protecting forests for the survival of communities. Sinha, Gururani, and Greenberg (1997) are critical of such representations. Asserting that Chipko was about the protection of forests purely for subsistence and protecting the traditional village unit overlooks the critiques of tradition, the calls for social justice, and the development aspirations that formed a major part of the movement. Indeed, as Guha (2000) relates, the initial impetus for the movement was not forest protection but employment. For several years, a local Sarvodaya collective in Chamoli district had been making requests to the Forest Department to gain access to local forests to establish small-scale timber cooperatives. After being repeatedly denied, the injustice of the situation became

movement has its origins after Gandhi's death, as an attempt to unite those engaged in 'constructive work' throughout the country. In the decades following Independence, Sarvodaya became an umbrella term for these decentralised Gandhian development initiatives throughout the country.

clear in 1973, when a large-scale commercial operation from the plains was granted access to a forest for which the Sarvodaya collective had unsuccessfully applied, just months before. To prevent the commercial foresters from taking what they saw as their own resources, villagers took to the forest to prevent the trees from being felled. At the suggestion of social activist Chandi Prasad Bhatt, they held the trees to prevent the loggers from being able to cut them—a gesture which would become emblematic of the movement.

From this time until the late 1970s, the movement spread throughout the region. Locals used direct-action tactics of embracing trees but also more 'militant' tactics, such as burning forests and timber auction houses (Sinha *et al*, 1997). Activists travelled from village to village to inform people of contemporary developments and methods of resistance (Weber, 1988). As Guha (2000: 179–184) relates, the movement was internally heterogeneous, with several variations in the analysis, goals, and tactics of various branches of Chipko. Guha suggests that different ideologies resonated more in different districts and regions. Kumaon's long history under British administration meant that the people were more exposed to 'modern' political ideologies, such as Marxism, and adopted a perspective more centred on local employment. In the district of Tehri-Garhwal, the charismatic leadership of Sunderlal Bahuguna played a crucial role. Bahuguna's ideas on a range of topics, including his Gandhism and his views on ecology, became influential in Tehri-Garhwal. These ideas had a strong influence on a cohort of young local activists, including those who would become the leading figures of BBA.

As the Chipko movement progressed, it was subject to a complex politics of representation. Guha (2000) suggests that Chipko developed two distinct identities: an inward identity as a peasant movement, focussed on forest access and livelihood concerns and an outer identity as an environmental movement, concerned with the conservation of forests. Rangan (2000) goes further, suggesting that the latter of these two identities constituted a 'mythic' representation of Chipko. She argues that Chipko's most prominent public spokesmen, Bahuguna and Bhatt, altered their discourse from one focussed on livelihood to one focussed on ecology to garner greater attention and traction in their engagements with media and politicians. This distortion became even more apparent when Chipko was represented to global environmentalist audiences—at which point the demands for local forestry and livelihood generation were almost completely obscured.

Such distortions within the Chipko narrative were not without consequence. Indeed, the Central Government's chief response to the Chipko agitations was to impose a ban on all forestry above a height of one thousand metres. This acknowledged Chipko's demands for environmental protection, but completely ignored the demand that local people should be able to establish small-scale forestry initiatives. Effectively, the ban reduced access to forests and increased the power of the Forest Department over people's lives—which was quite antithetical to Chipko's initial impetus (Baumann, 1998). Some have gone so far as to suggest that the continued demands for local development in subsequent social movements in the region demonstrate that Chipko's supposed environmental ethos was just an attempt to garner publicity and urban middle-class support (Baviskar, 2005). While myself and others have been critical of the suggestion that there was *no* grassroots support for Chipko's ecological demands (Brown, 2014a; Ishizaka, 2014), the dynamics of Chipko nonetheless demonstrate how subaltern perspectives can be sidelined within sustainability politics, particularly when activists become engaged with urban middle-class audiences.

These dynamics speak to matters of hegemony on two counts. First, they illustrate the hegemonic power of relatively privileged members of rural communities (Bahuguna, Bhatt) to speak 'on behalf of' the rural community as a whole. Second, and more fundamentally, they speak to the hegemonic power of dominant discourses, largely articulated by middle-class activists, which not only distort grassroots voices but also play a significant role in determining which issues and which voices reach the ears of the public and the state. Such hegemonic and discursive dynamics presented themselves once again in the case of BBA, albeit in a less pronounced manner.

The Chipko Origins of BBA

The history of BBA is intertwined with that of Chipko. Vijay Jardhari, Sudesha Devi, Dhoom Singh Negi, and Kunwar Prasun, BBA's leading figures, had all been involved in Sunderlal Bahuguna's Sarvodaya collective since the late 1960s. Its Gandhian ideals of village-level development have remained integral to their ideology. Indeed, in relating the history of the Chipko movement, these activists tend to place emphasis on the role played by Sarvodaya and Gandhians. In *Seeds for Life*, a short, English language booklet distributed by BBA that illustrates its history and philosophy, Dhoom Singh Negi and Biju Negi (n.d.) provide a genealogy of protest in Garhwal that highlights the influence of the Independence Movement—particularly the role of Gandhi. They note that Gandhism was highly influential after Independence, emphasising the role of the Sarvodaya Movement in redistributing land and combating the practice of untouchability.

In the early 1970s, when the Chipko movement was just starting, Vijay Jardhari, the founder and leading figure of BBA, owned a bookstore in the small town of Chamba, which sits above the Henwal Valley. At the time, Kunwar Prasun was a college student and Dhoom Singh Negi was working as a teacher in a nearby school. Jardhari's bookstore was something of an intellectual meeting ground and the three used to regularly meet to discuss local social issues. Being influenced by the Sarvodaya movement, they were particularly interested in discussing matters related to local community development. As Dhoom Singh Negi described:

> I was involved in Gandhian philosophy and social activities. Because we people in this area have very low income. In this way, I recognised that development … should not be only a single man's development. The whole society should be developed. This is Sarvodaya.

In their discussions, they considered methods of improving people's welfare through local action. They were aware of the negative effect that liquor had on local communities and became a part of the *Sharaab Bandi Andolan* (Hindi: Liquor Ban Movement) to oppose the sale of alcohol.[35] They held pickets along with local people, particularly women, who saw the adverse impact of men consuming all of their families' surplus income on liquor. When the Chipko movement became active in the region, the anti-liquor campaign resonated with the Gandhian perspectives of some of the Chipko activists. Sunderlal Bahuguna endorsed the demands of the anti-liquor movement, holding a *satyagraha*[36] in the city of Tehri to ban the sale of illegal liquor. Jardhari, Negi, and Prasun marched to Tehri in solidarity. According to Negi and Negi (n.d.), this was the beginning of a meaningful relationship between Bahuguna's branch of the Chipko movement and the Henwal-based group, stating that '[t]hereafter, the [Henwal Valley] group soon became among the keenest followers and confidante of Bahuguna' (Negi and Negi, n.d.: 22).

This emergent collective subsequently became involved in agitating for the protection of forests in the region. It was in this way that Sudesha Devi became a major activist in the Chipko movement. She recalls Bahuguna, Prasun, Jardhari, and Negi coming to her village and explaining the work that contractors wanted to do in the nearby forest of Advani.[37] She recalls the activists explaining the value of

---

35 Opposition to the sale and consumption of liquor has a long history in the hills. The issue was taken up on a large scale by Sarvodaya activists in the 1960s, who opposed it on moral grounds (Pathak, 1985). Within Gandhian thought, the consumption of liquor poses a barrier to the social, moral and economic upliftment of the nation (Gandhi, 1948).

36 Satyagraha is a term coined by Gandhi, literally meaning 'truth force', which refers to his methods of non-violent resistance and civil disobedience. They expanded their awareness of the inter-relations between social and environmental problems. By the mid-1970s, as Chipko began to gain momentum, the trio gave up their professions and became full-time Chipko activists. In his account of the Chipko movement, Guha (2000: 161–166) makes note of the trio's role in travelling from village to village, educating people about loggers' intentions and rallying support.

37 The forest at Advani was one of the major sites targeted for sustained direct action by the Chipko movement (Guha, 2000). It is within ten kilometres of the Henwal valley.

protecting forest, given the important role of forests in village life. Sudesha Devi cites the value of trees, along with the fact that the contractors employed to cut the forest were not local, as major reasons for her involvement in the movement. She became an active agitator within Chipko and rallied people within her village to become a part of the protest movement. In the initial stages, they occupied the forest at Advani and held the trees to protect them from loggers. She and other women from the village tied *rakhis*[38] to the trees to highlight their close relationship with them. After several days in the forest, they travelled to Narendranagar, where the timber was being sold. She and fourteen other members of her village occupied the auction hall, where they were confronted by hundreds of police. The fifteen people from her village were arrested and kept in gaol for fifteen days.

Partly through these efforts, the Henwal Valley became an important site of the movement, with two major incidents of villagers occupying forests and successfully expelling loggers. Bharat Dogra (2000), a journalist who has written extensively on Chipko and its later avatars, argues that the Henwal Valley was also significant for having placed particular emphasis on ecological issues. Indeed, one of Chipko's most famous ecological slogans was composed by Kunwar Prasun:

Kya hain jangal ke upkaar,	What gifts does the forest bestow?
Mitti, paani aur bayaar;	Soil, water and pure air,
Mitti, paani aur bayaar,	Soil, water and pure air,
Zinda rehne ke adhaar	The basis of survival.

Thus, although some scholars have suggested that Chipko was more a struggle for livelihood than the protection of local ecology (Baumann, 1998; Rangan, 2000), for the grassroots activists who would go on to become leading figures within BBA, livelihoods and ecology were clearly seen to be mutually dependent.

From Tree Hugging to Seed Saving

When the Chipko movement began to subside in the early 1980s, Jardhari, Prasun, and Negi decided not to return to their professional lives in the town but rather to their home villages. Following the Sarvodaya ethos, they would make their local villages the focus of their social action and take up the traditional source of livelihood—farming. During an interview, Jardhari related how he was particularly surprised to find the inroads that Green Revolution technology had made in his upland village of Jardhargaon. He explained that many people in Jardhargaon

[38] Rakhis are red threads which sisters tie around the wrists of their brothers on the Hindu festival of Raksha Bandhan.

and nearby Nagani (a largely irrigated village by the Henwal river) were using hybrid seeds and chemical fertilisers. Jardhari was alarmed by this development in village agriculture, feeling that the use of chemicals would cause damage to soils. Uncertain of the role of hybrid seeds in the equation, he conducted an experiment. In his first year of farming following his return, he used hybrid seeds and chemical inputs on his field. He found that their yields were approximately one and a half times greater than that produced by way of traditional varieties. In the second year, however, he tried to use the hybrid seeds without the use of chemical inputs and found the yields were very low. He realised the dependency of hybrid seeds on chemicals and that, therefore, for non-chemical agriculture to continue to thrive in the region, it would be necessary to return to the use of traditional seeds.

Talking with other villagers, Jardhari found that his concern regarding the impact of chemicals on soil was met with a similar concern regarding the loss of traditional seeds as the use of hybrids became more widespread:

> So we started to feel that our fields and our crops had become like drug addicts … Without the chemicals, they can't grow. So we went and spoke with the people in the village, and said that we feel that this is damaging our soil, and they told us that it's not only that—we used to have so many varieties of seeds… but now that's all finished.

Jardhari reported that during the time immediately following the conclusion of the Chipko movement, hybrid seeds were being used in both irrigated fields in the valley, and in the unirrigated fields, further upland. He suggested that scientists and the Agriculture Department had been promoting hybrid seeds in the villages, particularly a hybrid variety of soy bean. Jardhari said, however, that the use of soy bean created considerable discontent, since although those who adopted the seeds during the initial phase enjoyed a good harvest, by the second season, the sheer number of people growing the crop in the region led to over-supply and farmers struggled to sell their produce. When Jardhari and Sudesha held meetings with local village women, they found that the women were also experiencing problems with the soy bean crop, as unlike wheat and rice, its straw could not be used as fodder. Furthermore, it was not possible to produce oil or milk from the soy bean at the village level, as this would require a more industrial operation, which was neither feasible nor suitable to the village economy. Thus, the use of soy bean to meet basic subsistence needs was quite limited, when compared to traditional crops.

For Jardhari, it was evident that a return to the traditional approach to farming was necessary. When he tried to find traditional seeds within his own village, however, he could only find two varieties of rice in use. Thus, in 1985, he mobilised his friends and contacts from the Chipko movement to form the Beej Bachao

Abhiyan (Save the Seeds Campaign), as an attempt to re-establish the genetic diversity of the region's agriculture. Members of the campaign travelled to the more remote villages of Uttarakhand, which the Green Revolution technologies had been unable to reach, to collect indigenous seeds. Between 1985 and 1987, they collected hundreds of traditional varieties of wheat, rice, millet, and rajma (kidney bean). When visiting villages, they encouraged people to be on the lookout for local varieties of seed, to make an effort to conserve them, and, if possible, to pass some on to Jardhari, so they could be stored in a collective seed bank. The seeds stored in the seed bank were then made available to anyone in the region who wanted to return to traditional farming systems. In the years that followed, BBA travelled from village to village to distribute them.

Image 8: Some of the diverse varieties of rajma (kidney bean) amassed in BBA's seed bank

After several years of operating in this manner and having influenced many locals to return to traditional seeds and traditional agriculture, the participants began to conceive of their initiative as more than merely a 'campaign'—it had become a movement. At the instigation of members, therefore, Jardhari renamed the group as Beej Bachao Andolan (Save the Seeds Movement). In referring to their initiative as a 'movement', BBA activists asserted that their work was part of a more permanent phenomenon with a more holistic focus. Furthermore, through using the term *andolan* (movement) to describe themselves, Jardhari and others situated themselves within Garhwal's tradition of social movements. Indeed, the various andolans in the region were seen by many participants to be of one ethos and part of one continuous phenomenon. Dhoom Singh Negi, for example, highlighted the continuity between BBA and the Chipko movement and the movement against the Tehri Dam:

> They are not different—Anti-Tehri Dam Movement, Beej Bachao Movement, Chipko movement—they are of the same thinking, same strategy, same planning, future planning, [the same] concept of development … Save environment, save nature and save the villages, save communities and save humanity. There is one idea in all [these movements]: what is development? The existing development theory is wrong. There is destruction in it. Co-operation with nature—this is the real development.

Thus, all of the local movements are seen to be related to the environment and also to the protection the village economy and local livelihoods. Having said this, however, some participants did acknowledge differences between BBA and previous movements in the region. Gautam, a BBA activist, noted that where the other movements had a focus on singular, short-term issues, BBA represented a more continuous process:

> Whatever other movements there were, they were all about one issue. That is, after the cutting of the forests is done, the forest movement is over. There was the [Uttarakhand] State Movement—now we have gained our statehood, so it is over. But Beej Bachao Andolan is about the necessities of life and so is a continuous movement, just as the necessities of life—the sowing of seeds, the raising and feeding of crops—are a part of the continuous process of life.

Because of their focus on the more quotidian aspects of village life, BBA proceeds at a slower pace than the other prominent movements in the region. In an interview, Sudesha Devi brought out this point, observing that 'There's not so much agitation; we are just trying to make people understand. Meetings, going from village to village for meetings'. The Chipko movement, she says, was 'hard work', with direct conflicts with police and other authorities. Activism in BBA has been more peaceful.

Though BBA activists do not overtly contrast their form of organisation to that of NGOs (in the manner of KVM), it is clear that they are uninterested in adopting a typical NGO structure. Some of the group's early experiences with funding made them particularly cautious of the NGO approach to promoting sustainable agriculture. Jardhari related that, in the late 1980s, he was visited on several occasions by a prominent Indian environmentalist (here unnamed, for confidentiality purposes), who was impressed with the diversity of seeds he had managed to collect. The environmentalist praised the effort and suggested that Jardhari should receive funding to bring those seeds to other villages. She provided some funds for Jardhari to carry out this work, having told him that the money was 'from friends'. After carrying out this work for some time, Jardhari was surprised to find that the funding was actually coming from the United Nations' Food and Agriculture Organisation and was a part of a larger project. Jardhari emphasises that he is not opposed to the receipt of funds *per se* but was concerned that this money had come from an organisation of which he himself was critical. He felt betrayed by the environmentalist's dishonesty in not disclosing this to him.

> We said to her ... 'We are not opposing any project, but if we are a part of this project, then it should have been an open project amongst us, open amongst all of our friends regarding who is running the project'.

Furthermore, it was found that when reporting on BBA in the media, the environmentalist was referring to BBA activists as 'my activists'. Jardhari and Sudesha Devi both expressed their frustration at this comment, which seemed to imply that the environmentalist was claiming ownership of their work. They thus decided that this project was not in their interests and that in future they should operate without receiving funds for projects. Since that time, BBA has worked more or less autonomously and without receiving any major grants. As such, they would appear to be less subject to the hegemonic influence of donor organisations than most sustainable farming initiatives operating in India today. The work that they do generally reflects their own conscience, rather than donor priorities. Costs involved in hosting meetings and workshops, travel, and so on, are covered by activists themselves, through their activities as farmers, writers, and researchers. More recently, BBA has managed to bring in some additional income through the sale of their organic produce and handicrafts at the Spirit of Uttarakhand festival, held annually at Dilli Haat, New Delhi. On certain occasions, activists are invited to give lectures and workshops in other parts of the country and for these purposes may be given financial support by like-minded organisations.

BBA can also be distinguished from an NGO in terms of its formal structure. In the past, BBA was more focussed on its central task of collecting and storing seeds. Now that a large seed collection has been amassed, this is not so pressing and so the movement operates at a slower pace. Activists described the movement as 'informal' and 'fluid'. Meetings are called as the need arises, such as when a new threat emerges that needs to be addressed. To maintain some sense of continuity, however, participants try to make contact at least once every two months or so. They also participate in the Mitra Mela (Friends' Festival), an annual convergence of Uttarakhand's activists. The Mela is a key medium through which activists are kept up to date with contemporary local issues and struggles. BBA does not have a fixed membership structure and can be seen more as a network of supportive contacts with a mutual willingness to collaborate when some activity needs to be coordinated. One activist suggested that this has been both a blessing and a curse for BBA. It allows flexibility, but the lack of formal structure and institutionalised power means that the movement is unable to exert strong pressure on people to change—its members can only lead by example. Furthermore, the lack of funds means that BBA is unable to develop projects to counter the encroachments of Green Revolution style agriculture throughout the region (many of which are sponsored by agricultural universities and other large institutions). This institutional weakness of informal groups, such as BBA, relative to sponsored NGOs, is one of the ways in which NGOs come to exert a hegemonic influence over the development process (Kamat, 2002).

The Philosophy of BBA

As its name suggests, BBA is in the first instance defensive. It aims to 'save' the diverse varieties of seeds that are indigenous to the Himalaya from the onslaught of corporate agriculture and opposes the use of hybrid seeds.[39] BBA activists cite a number of reasons for this opposition. First, they argue that hybrid seeds are inappropriate for the social, ecological, and climatic conditions of the hills. In an interview, Sudesha Devi made several points to illustrate this. She pointed out that hybrid seeds are expensive and require farmers to purchase them year after year. They are also dependent on chemical fertilisers, which is a further expense. Farmers in the region cannot afford such costs. Furthermore, she asserted that the use of chemical fertilisers damages the soil, whereas the manure applied in traditional agriculture improves soil and allows it to hold moisture. Since the region's agriculture depends heavily on rainfall, which is often unreliable, maintaining the soil's water-holding capacity is particularly important. Sudesha also noted that traditional varieties provide a superior source of fodder, especially when compared to the hybrid soy beans that were being promoted in the region. In addition to these issues, Gautam mentioned that the supply of hybrid seeds to

[39] In interviews, both activists and farmers used the terms prakritik (natural), paramparagat (traditional), or purane (old) to describe traditional seeds and vigyanik (scientific) to describe hybrid variety seeds. Occasionally, the distinction was between ghar ke beej (seeds from home) and bazaar ke beej (seeds from the market).

the hills was unreliable, thus making dependency upon them risky, particularly if traditional seeds were lost in the process. None of the other farmers interviewed for this research used hybrid seeds. When asked why they refrained from buying hybrid seeds, they cited the excessive costs and/or a view that such seeds would not be effective in the hills.

A second reason for BBA's opposition to hybrid seeds is that they displace traditional varieties. Jardhari emphasised that the traditional seeds of the region are grown in diverse cropping systems, which can better meet the nutritional requirements of households than hybrid varieties, which are grown in monocultures. As agriculture in Uttarakhand is mostly subsistence-oriented, this is an important point. Furthermore, he argues that a loss of diversity makes the farming system of the region more susceptible to failure in the event of environmental changes, such as global warming. BBA activists speak of the importance of the traditional system of farming in the hills, known as *baranaja* (from Garhwali, *barah anaj*, 'twelve grains'). This mixed cropping system is today mainly used in rain fed areas during the *kharif* season. It involves sowing up to twelve different varieties of seeds in a single plot, each of which is harvested at a different time through the season. The various seeds planted have complementary growth and nutritional cycles, making the system quite productive. BBA highlights the importance of *baranaja* as a system that brings a continuous flow of diverse food into the home (as each variety sown is harvested at a different time), optimises the use of land, and minimises the risk of crop failure. Smriti, a BBA activist who had promoted *baranaja* at several academic conferences, explains some of the chief benefits of the system, given the extreme variations in local climate:

> When there is a drought, the chaulai and mandua [varieties of amaranth and millet grown in the baranaja system] have very good production … And in the opposite kind of weather, when there are floods, there are some grains that do well … So no matter what happens our people are able to save some food.

By contrast, when mono-cropping with hybrid varieties, variations in weather risk the loss of an entire crop.

A third reason that BBA opposes hybrid seeds is their belief that neither corporations nor any individual has the *right* to sell seeds, since seeds were a gift of nature which farmers have bred over thousands of years. In this respect, in particular, BBA articulated a philosophy that directly challenges capitalist ideologies and the hegemony of agri-business over third-world farming systems. Negi and Negi (n.d.) describe the process by which farmers traditionally kept aside the best section of a crop, and rather than consuming it, used its seed for the following year. In this way, crops were bred to be more productive over time.

The manufacture of hybrid seeds involves taking these very seeds which farmers have improved over millennia and making relatively minor manipulations to make them more receptive to chemical inputs. According to BBA, seeds cannot be seen as the property of the scientist who manipulated them or the corporation that sells them; they are the collective property of humanity and should be guarded as such. For Charan, a BBA supporter, corporate ownership of seeds is a great danger to the welfare of farmers, as it represents a loss of control over their natural heritage and a conferring of all rights to agricultural scientists. He suggests that farmers should be recognised for their scientific innovation in developing seeds, and that whatever scientific experiments are done to improve seeds should be done by farmers themselves in a manner suitable to their local environment.

While having this strongly defensive dimension, BBA's philosophy should not be seen as simply a reaction to developments outside of its own control. Their claims that hybrid seeds and other Green Revolution technologies are inappropriate for village life are rooted in a Gandhian view of village economies and rural development, which was propagated in the Henwal Valley with the Sarvodaya collectives and later by the Chipko movement. They hold to the ideal that villages must develop from the ground up and in a manner that is suitable to local culture. To this end, control over seeds serves as a metaphor for controlling the trajectory of local development—albeit a metaphor with great practical significance, as seeds are central to the reproduction of agriculture and losing control over seeds takes village development out of the hands of locals. As Jardhari explained:

> Beej Bachao Andolan's philosophy is this—the freedom and self-respect of farmers. And first of all, this means agriculture—this is our livelihood, nature and culture [*jeevan, prakriti aur sanskriti*] … In India we believe in *Dharti Maa* [Mother Earth], and that we have a moral obligation [*vyohaar*] towards her. It's like a child who drinks its mother's milk, and will continue to drink it for six months or, if necessary, up to two years, but with the Green Revolution philosophy they want the mother to continue to produce the maximum amount of milk. If we continue to artificially take the maximum amount of milk it will be bad for both mother and child. The mother's health will be damaged and if the child continues to take all of the milk then what will happen? So the relationship between nature and human beings should be like that between child and mother. So we believe it's our responsibility to take care of the soil as our mother, so that she will be able to provide not just for ourselves, but also future generations. That's why we want to get rid of the chemicals and return to the natural way of doing agriculture, because we saw the violence [*himsa*] of the Green Revolution. Where the earth had sustained life, these chemicals will destroy it … Our traditional agriculture is, in a way, non-violent agriculture. And if we continue to treat the soil as a child treats its mother and use our traditional seeds, we believe that will provide all of

us with our sustenance for a long time. We follow a Gandhian line of thought, and Mahatma Gandhi ji once said a very beautiful thing: that the earth can provide everyone with their sustenance [*poshan*], will give them their food … everything they need for living. But if we think about greed, [the earth] can't give enough to satisfy even one man.

BBA's philosophy is thus more holistic than its name might suggest. BBA promotes a disposition of custodianship over land, with a view that takes the survival of future generations and the village unit as central priorities. In their philosophy, at least, they are therefore not only counter-hegemonic but also articulate a vision for an alternative based on autonomy that has much in common with discourses of 'food sovereignty'—a term which they invoke in their publications. As Negi and Negi (n.d.: 27) outline:

… Beej Bachao Andolan is not just about saving seeds. It is not even a task or an activity. It is a thought. It is a way of life. At the centre of its philosophy is the balanced web of life that includes the five elements—earth, sky, water, air, and fire—that encompass creation. It is about humans, but as much about animals, insects and the plants. It is about equity and equality, about social coherence and togetherness. It is about food security and food sovereignty. Beej Bachao Andolan is about the survival of our being and about the onus of this on man as the caretaker.

Beyond the concrete practices of seed-saving, BBA are defending an essentially Gandhian view of development. Their focus on the productivity of local subsistence agriculture attempts to emphasise that development should not be exclusively about economic growth and access to an increasing range of modern commodities—a development model based on greed, which they see as unsustainable. Although they are not opposed to development *per se* (Jardhari emphasises that he endorses the increased access to health and education facilities that have come with modernity), BBA's model of development places greater emphasis on increasing cooperation and better meeting local subsistence requirements.

BBA's discourse is thus informed by (a) the long-standing influence of Gandhian thought in the region and on its leading activists in particular; (b) the ideas that developed during the Chipko agitations and subsequent environmental movements regarding the importance of ecology and local subsistence farming systems; and (c) their meaningful grassroots engagements with the realities of village life and the needs of the subsistence economy. In this sense, BBA has many 'grassroots' features that differentiate it from other sustainable agriculture initiatives in India. Rather than being informed by the hegemonic interests of the state or donor organisations, it is a response to contemporary agrarian issues that emerge from the perspective of small-holding farmers. Furthermore, given that the local agrarian structure is relatively flat, the potential for co-optation by local elites was never a serious concern. Yet, despite this strong connection to the grassroots, BBA has suffered a decline in local relevance in recent years, the reasons for which require careful consideration.

Impact: Local, National, and International

BBA's story of grassroots seed conservation and defence of traditional agriculture has been an inspiration within the sustainable agriculture community in India and, indeed, internationally. Yet, while BBA's message has found numerous supporters in national and international activist circles, its status among farmers in the Henwal Valley is more ambiguous. While hybrid seeds are mostly rejected, BBA is not seen as having solutions to more immediate and pressing local challenges.

During the late 1980s and early 1990s, BBA activists held numerous meetings with local people to spread their message. They made use of existing organisational structures within villages, such as Mahila Mandal Dals (women's groups), to mobilise local people and organise meetings. Dogra (2000) suggests that women were initially more receptive to BBA's ideas. Because of their more prominent roles in subsistence agriculture and village life, Dogra argues, they recognised the risks involved in losing control over seeds, whereas men, who traditionally had the task of buying goods from the market, were more directly influenced by seed sellers. With time, however, men's perspectives also changed. The impact of this initial campaign can be seen through current agricultural practices in the region. Of all the farmers interviewed for this research, none were using hybrid seeds, citing both cost and a belief that they would not be effective as reasons for their rejection. Several farmers said that they are occasionally confronted by people from seed companies, NGOs, or the Agriculture Department, who try to convince them to use hybrid seeds, but they remain sceptical. While the majority did not use chemical fertilisers, some indicated that they occasionally used urea. Only one participant acknowledged using chemical pesticides at times when her crop appeared unhealthy. Thus, while the majority of farmers interviewed were farming mostly according to traditional systems, it could not be said that chemical use is entirely absent from the region.

BBA's local influence extends beyond farmers—its ideas and values have also been taken up by a number of local development organisations. In an interview, the leader of one local NGO noted that unlike other development NGOs, which had tried to impose often inappropriate development models from the top down, BBA's philosophy emphasised the value inherent in tradition. His NGO drew on Jardhari's writings on *baranaja* and his ideas relating to seed saving to formulate development projects. Another local NGO worker and activist suggested that of all the movements that had occurred in the Henwal Valley, BBA was the most important, as it directly spoke to issues of livelihood, which have direct relevance for villagers:

> This is the main issue. This is the livelihood issue. This is the sustainability issue … They are fighting for the farmers. Many times we will go to the village and women will say to us, 'We do not want anything, we just want our agriculture. Save our agriculture—from wild animals, save our agriculture from climate change'.

Inspired by BBA's emphasis on agriculture, this NGO worker had attempted to use agriculture to develop sustainable employment opportunities for local youth. He was particularly interested in developing small-scale industries that allow value adding for local crops and forest produce.

BBA's story and philosophy also have a special significance within the sustainable farming movement throughout India. The chapter on organic seed in Alvares' (2009) *Organic Farming Sourcebook* opens with a three-page interview with Jardhari. It is clear that Jardhari's story has a special resonance. It brings the binaries that are used within the sustainable farming movement into sharp relief—for example, traditional versus modern farming, communal versus corporate ownership, and biodiversity versus monocultures. BBA makes a relatively unambiguous case for the inappropriateness of chemically intensive, corporate agriculture, since the absence of a clear class of capitalist farmers in Garhwal means that there would be few clear beneficiaries. It thus validates sustainable agriculture discourses, which often emphasise the value of subsistence systems.

Because of this broad discursive resonance, BBA has become well networked within sustainable agriculture movements, in India and abroad. Jardhari and other BBA activists are frequently invited to give seminars throughout India on their various initiatives and Garhwal's traditional agriculture. They have also attended several international conferences. Building on their connections within the Pesticide Action Network (Asia-Pacific), BBA activists have attended seminars in Penang, Malaysia. Jardhari and other BBA members have spoken at conferences there on topics, such as 'Food Crisis and Climate Change', where BBA was seen as a case of local resilience in the face of global ecological turmoil. At other events in Penang, BBA members have spoken on the value of *baranaja* and other mixed cropping systems for upland agriculture in Asia. At the time fieldwork was conducted, participation in these national and international events was taking up a large portion of Jardhari's time.

While enjoying this general appeal in activist circles, the key challenge for BBA is sustaining the relevance of its core message within the Henwal Valley. Although

BBA activists have continued to be involved in various local campaigns (see below), in recent years, they have done little work locally to propagate their philosophy or to build more sustainable farming systems to support local livelihoods. Indeed, in interviews conducted with local people, there did not appear to be any great enthusiasm relating to the issues raised by BBA. When asked about their views on BBA, most farmers (except those directly involved in the movement) responded that they knew of them but found their message somewhat underwhelming. In some respects, this challenge of sustaining relevance may be attributed to the very 'success' of BBA. The use of traditional seeds and farming methods in the villages of the Henwal Valley has become common sense and needs no further emphasis. Indeed, as several activists noted in interviews, even the agricultural universities and other state institutions, which were formerly hostile to BBA's message, have begun to adopt the view that Green Revolution style agriculture is not appropriate for the region. Since there is no strong push for an alternative system ('Green Revolution'), the very notion of the 'traditional system' becomes more blurred and less politicised. Much of the impetus of BBA is exhausted, at least at the local level.

The decline in relevance is reflected in the difficulty BBA has faced in recruiting young members. Most of BBA's more active members are veterans of older struggles, such as Chipko, the Tehri Dam movement, and anti-mining campaigns. Young people do not seem to have found compelling reasons to become involved in BBA. The few young activists who have tried to apply BBA's message have done so via other institutions. One interviewee, who worked at the Uttarakhand Organic Commodity Board, suggested that the lack of young participants in BBA was preventing it from considering alternative means of using agriculture as a means of livelihood generation in the contemporary setting, such as gaining premiums through organic certification:

> V: Beej Bachao Andolan until now has not been able to bring in its farmers for certification, but we've [the Organic Commodity Board] always told them this government is giving a one hundred per cent subsidy—it's free, actually, under the certification regime. We've got a big budget for it and Beej Bachao Andolan farmers have been told through Jardhari ji many times that they must come and access the services.
>
> TB: So is there some barrier that's stopping them from coming?
>
> V: I don't know, there is no barrier, really. What I feel now is that Beej Bachao Andolan, I think it needs some more younger people in the group, you know, for the movement. I think Jardhari is getting old and he is too busy in many other sectors—because it's not just agriculture—it's anti-mining, it's legal issues and other things.

Aware of the issue of declining participation, Jardhari said in an interview that it is difficult to change the attitudes of youth through this movement and change would need to come through education. However, he then added that by focussing more on local employment generation, such as fruit cultivation and small-scale food processing, they could appeal more to youth. At the time the fieldwork was conducted, however, there was little evidence that BBA was taking up these options. It seemed that their time was spent responding to opportunities within national and international sustainable agriculture networks, more than to the calls for livelihood generation emerging at the grassroots. In effect, the desire for self-validation within these predominantly middle-class activist networks—validation of their assumptions on the merits of non-chemical and traditional farming—created a structure of opportunities for BBA activists, which syphoned them away from their grassroots constituents.

Interviews with local people showed that there are a number of pressing issues in the region for which BBA appears to provide no answers. Perhaps the most serious of these is climate change. All farmers interviewed commented on diminishing rainfall in recent years and corresponding crop failures. Indeed, this was generally the first and most emphatic issue that they spoke of when discussing agriculture. BBA activists have used climate change as an entry point to discuss the value of traditional millet varieties, which are more drought-tolerant than wheat. Yet, several farmers claimed that in recent years even the millet crops have performed poorly. BBA could conceivably be more responsive to this threat to the sustainability of local agriculture by focussing more on water-conservation strategies. For example, they could take up scholars' recommendations of adopting water management programmes and construction of small dams and tanks to overcome water shortages (Krishna, 2002). Indeed, other local NGOs had been focussing on cropping methods that conserve water, such as the System of Rice Intensification (SRI). For an unfunded and loosely structured group, such as BBA, however, it is difficult to mobilise resources for such initiatives.

A further problem affecting agriculture was the accelerated out-migration from villages. The inability of the current agricultural system to meet basic subsistence requirements has led more people to leave villages in search of work in cities, in many cases abandoning agriculture altogether. It is not uncommon for the entire working age male population of villages in Tehri–Garwal to be absent for much of the year. While the seasonal migration of men from villages has been commonplace in the region for a considerable time (Bora, 1996), anecdotal accounts suggested that women are now moving with their husbands, rather than staying in the village and working the fields. When this occurs, migration tends to be permanent, rather than seasonal. Reportedly, this process has accelerated. Some farmers noted that in their villages, many fields have been abandoned. When asked about the future, villagers were profoundly pessimistic. One, for example, suggested that within a decade everyone will have either left or died of starvation. The pronounced out-migration suggests that many villagers do not see the local, village economy as being capable of meeting their needs—contrary to BBA's Gandhian message. Indeed, while local people have accepted BBA's message that Green Revolution agriculture is not suited to local economic and ecological conditions, this does not mean that they have accepted their Gandhian philosophy of moderation and serving as good environmental custodians. Rather, they pursue their developmental aspirations by migrating in search of cash income, rather than through agriculture.

The issue of outmigration highlights that the terrain of the struggle has shifted since the time of BBA's inception. If the objective is to present self-sufficient rural communities as a viable alternative to the hegemonic food regime, one must begin by acknowledging that in remote places, such as Uttarakhand, the hegemonic mechanisms of consent and coercion are no longer corralling people towards Green Revolution style agriculture with the kind of insistence evidenced in previous decades. Instead, they appear to be (quite effectively) pushing for an exodus from villages altogether, whether to promote the consolidation of land holdings in the villages or—more importantly—the consolidation of a 'reserve army of labour' in the cities. While the self-validating mechanisms of national and international sustainable agriculture networks have led to a partial ossification of BBA's message as 'anti-Green Revolution', an opportunity to present a more contemporary counter-hegemonic vision of Gandhian sustainable rural development appears to have been missed.

Sustaining a Tradition of Activism

While BBA may have become less involved in constructive work for livelihood generation, its activists have continued to participate—and often play leading roles—in local campaigns against threats to the integrity and sustainability of local agriculture. While working for these causes, Jardhari and others may not use the name BBA, but they continue to draw on its philosophy. Their participation in the campaign against the construction of the Tehri Dam in the 1990s, based partly on their concerns regarding its effects on local water flows—and subsequently on forests, irrigation, and agriculture—testifies to this. BBA activists have also been key participants in local campaigns within the Henwal Valley itself. This is clearly demonstrated by their involvement in local campaigns against mining, the construction of large power lines, and the encroachment of wild animals on agriculture.

In the late 1990s, a contractor was given a licence to mine limestone in a location very close to the village of Kataldi in the Henwal Valley. When operations started, the lives of local people were seriously disrupted by constant explosions. Flying debris resulted in damage to property and injuries to both livestock and people. Women were unable to continue with their farm work during the day, as it was unsafe to venture outside while blasting was occurring. Furthermore, it was alleged that the blasting had disrupted underground water flows (Kohli, 2004). One local woman, Geeta, initiated a campaign through the local Mahila Mandal Dal. Jardhari attended and participated from the first meeting. BBA activists provided moral support and tactical advice to the women of Kataldi, telling them that they could win, citing their own victories in the Chipko movement. BBA's participation was important, as there was a division at the local level between the village of Kataldi, where the mine was causing disruptions, and other nearby villages, where the mine was seen as an employment opportunity. BBA activists encouraged people from villages outside Kataldi to support the Kataldi campaign. They prompted women from other Mahila Mandal Dals to become involved in the campaign, on the grounds that agriculture is the main source of livelihood for women in the region and anything that threatens agriculture is a threat to all of them. After using direct action tactics and holding several rallies, the campaign was successful, with a court ruling in favour of the Kataldi villagers.

In 2001, BBA assisted in mobilising opposition to a large development at Advani village. There was a proposal to construct large power lines near Advani, which would entail clearing the forest eighty-five feet on either side of the lines. The lines would pass directly through the forest at Advani, meaning reduced access for villagers and a significant loss of forest cover. This was particularly significant, as Advani was the site of one of the largest direct actions of the Chipko movement. BBA members, among others, mobilised popular anger that the forests, which had been protected by their work in Chipko, could be taken away from them so easily. BBA helped villagers to lobby government departments and threatened to repeat Chipko's direct action tactics. Ultimately, the campaign was unable to prevent the construction of the power lines, but the area of forest felled and the reduction in forest access rights were both greatly reduced in response to their demands.

More recently, BBA has mobilised in response to local concerns regarding an increase in the number of wild animals in the region. Almost all farmers interviewed for this research identified encroachment by wild animals, which cause damage to crops, as one of the main issues affecting agriculture in the region. According to Jardhari, animals that had been displaced by the Tehri Dam were unable to sustain themselves on local forests and were forced onto farmland to consume crops. In 2009, Jardhari and others organised a rally at the village of Nagani in the Henwal Valley, to demand that the government address the animal problem. Locals demanded the construction of electric fences and the employment of a watchman to shoot pigs that encroached on the land. Drawing on their Gandhian ideology, however, BBA activists also made a 'non-violent' suggestion that more trees be planted in the forest that could serve as appropriate food for the animals and prevent their encroachment on fields. According to Jardhari, the government agreed to these demands, but it was unclear whether they were implemented.

BBA is thus involved in sustaining a tradition of activism that is engaged with a range of issues of relevance to the local village economy and society. It draws awareness to issues affecting local subsistence agriculture and, in doing so, highlights the ongoing importance of these farming systems and the local ecology to people's lives. In all these cases, however, BBA merely resists the forces unleashed on villages as a result of the ongoing capitalist development of the region. Since its core message surrounding seed conservation and the value of traditional farming systems is largely exhausted, it can no longer be regarded as actively building the autonomy and sustainability of local farming systems. Without this, it cannot stem the flow of out-migration. Indeed, it seems that the more BBA's core message is made to resonate with national and international activist circles (highlighting the 'wisdom' of farming traditions, such as *baranaja*), the less resonance it has at the grassroots as a truly counter-hegemonic alternative. In the 1980s and 1990s, it may have successfully resisted the corporate encroachment into agriculture; but in the present, it has not succeeded in resisting the broader neoliberal hegemonic project of draining people from the countryside to constitute a new reserve army of labour in the cities. Sustainable farming systems may be the preferred alternative to Green Revolution agriculture in Uttarakhand but they are very far from the preferred alternative to urban-based employment.

Conclusion

In several respects, BBA differs from the previous two case studies presented in this book. Whereas in the previous cases, the leading figures within the organisations were neither from farming backgrounds nor local to the villages in which they work, BBA is led by people from the villages in which it is active. Although Jardhari, Prasun, and Negi were well educated and thus may be construed as having greater cultural capital than the 'average' local, as Gandhian activists, they had made the principled decision to work in agriculture as the traditional form of livelihood in their communities. This, they believed, would keep them sensitive to local issues. Furthermore, BBA, unlike the other organisations studied here, does not engage in project work nor receive institutional funding. Their organisational structure is largely informal and their work focusses on raising awareness about issues that are important to the region's agriculture and lifestyle and agitating to defend rural society from what they perceive to be destructive encroachments from outside.

Their local roots and their refusal to engage with mainstream funding sources make them removed from traditional hegemonic forces and a valuable case study of grassroots opposition to the hegemony of corporate agriculture. Their seed conservation efforts have involved rural Uttarakhandis in the process of building their own self-sufficiency and less dependent on the market and the various local campaigns that they have supported have protected local subsistence systems.

The declining appeal of BBA's core message within villages, however, needs to be subject to careful consideration. As this chapter has demonstrated, while BBA's oppositional activities may have relevance at the grassroots, in the contemporary setting, their promotion of traditional agriculture does not generate a strong response in local communities. On the one hand, the essentials of its philosophy have become common sense. Local people generally share BBA's view that hybrid seeds are inappropriate for the hills, and therefore, the message of rejecting them does not strike most farmers as particularly profound. On the other hand, BBA's message fails to speak to the more urgent issues affecting agriculture in the region at present, particularly out-migration and climate change.

Understanding hegemony in contemporary rural Uttarakhand may require a shift in focus: hegemonic forces, rather than building consent for a shift to intensive commercial agriculture (clearly an exhausted project in the state, where the Government of Uttarakhand actively promotes organic farming), build consent for a complete exodus from agriculture. At this stage in history, those in precarious resource zones, such as Uttarakhand, are more valuable as an urban reserve army of labour than as participants in commercial agriculture in a region that would always struggle to produce surpluses. Through a myriad of capillary mechanisms (particularly within media and education), the state positions out-migration as the defining aspiration for Uttarakhand's rural youth. BBA lacks the resources and organisational structure to develop effective and coherent rural alternatives to out-migration. Building sustainable rural livelihoods that the next generation could aspire to would evidently require methods of farming that the traditional system does not provide, particularly while climate change and incursions by wild animals undermine the productivity of these systems. Grassroots resistance needs to evolve to keep up with this trend, and BBA's Gandhism makes it somewhat ambivalently disposed towards exploring livelihood options that youth would find acceptable, tending to descend into somewhat romanticised representations of the merits of traditional rural life.

A further issue, is the way in which BBA, despite being led by local activists, has gradually become entwined in non-local activist networks—both national and international. In recent years, BBA's core activists have found more receptive audiences for their message regarding seed conservation and the value of traditional systems through local, national, and international NGOs and social movements, rather than in the villages of the Henwal Valley. We need to seriously consider whether these networks, largely composed of urban middle-class activists, are framing their political discourse in terms of the genuine needs and aspirations of the rural poor or rather their own interests in preserving romanticised images of rural life and retaining 'purist' environmentalist discourses. Responding to the opportunities opened up within such activist circles inevitably shapes the direction of BBA's movement and ideology, guiding them towards messages that confirm the assumptions of sustainable agriculture movements in India and abroad. This may, regrettably, also amount to subtly guiding them away from engagement with grassroots concerns.

This is not to say that BBA activists themselves are irrelevant or detached. Indeed, as has been outlined in this chapter, the leaders of BBA have been highly involved in more contemporary local struggles, such as the opposition to the mine at Kataldi and campaigning for government assistance to resist incursions by wild animals. It may be said, however, that BBA has had difficulty responding to issues relating to the productivity and viability of agriculture in a changing ecological, social, and economic context. This may be attributed (1) to their lack of resources and organisational structure; (2) to their Gandhian ideology, which is somewhat ambivalent towards local development aspirations; and (3) their imbrication within middle-class activist networks, who would rather see them confirm their assumptions regarding the value of traditional subsistence agriculture than acknowledge that the traditional system may need to be modified to cope with a changing world.

7

Conclusion

Every crisis is an opportunity for change: an opportunity to rethink and reconfigure arrangements that have long since become defunct. A widespread agrarian crisis in rural India has provided an opportunity to rethink patterns of rural development that have failed to yield sustained benefits for most of the rural population. Insofar as this agrarian crisis has developed into a 'crisis of authority'—a weakening of the bonds of consent on which hegemony is based—it has offered a dual opportunity. For historically subaltern groups, it has been an opportunity to reconfigure power relations in a manner that runs counter to the established hegemony and to propose alternatives that might be more in their long-term interests. Yet, for hegemonic groups, it is an opportunity to further consolidate their power: capitalising on widespread desperation to push through reforms that facilitate further cycles of accumulation.

Throughout this book, I have attempted to assess whether sustainable agriculture initiatives in India tend to support a counter-hegemonic agenda of the rural subaltern or, alternatively, whether they align more with the existing hegemonic configuration. The case studies presented show that sustainable agriculture, though lauded as the solution to a pervasive agrarian crisis, is a project that, in many ways, is woven into the fabric of the same hegemonic structure that has produced the crisis in the first place. Though the ideology of sustainable agriculture initiatives often reflects its 'promise' as a pathway to greater farmer autonomy, a cornerstone of 'food sovereignty', and a challenge to corporate control in agriculture, the institutional context within which sustainable agriculture projects are implemented often results in this promise remaining unfulfilled. Because of their continued dependency on hegemonic groups, sustainable agriculture initiatives in the Indian context often deliver results that are either ineffective or further entrench existing power structures.

In this final chapter, I would like to suggest two major implications of these findings, which speak to two distinct audiences. For those within the development establishment, particularly those with an earnest commitment to delivering results that are effective, sustainable, and equitable, the findings suggest a need to rethink the best pathways to inclusive and sustainable development. For those outside of the development establishment, whether activists or academics, who are interested in more radical alternatives to the corporate food regime, the findings suggest a need to accept that, at least at present, sustainable agriculture does not offer a pathway to a genuine reconfiguration of power in most parts of India. I expand on each of these conclusions below. I begin, however, on a more optimistic note, by acknowledging the genuinely transformative and potentially counter-hegemonic moments (however small or transient) that were found within the case studies and their potential significance in creating viable alternatives to the corporate food regime.

Glimpses of Food Sovereignty

While the case studies presented here may not have fully lived up to the vision of 'food sovereignty' or a grassroots, farmer-led sustainable agriculture movement, they all did, in various ways, provide glimpses of what such a movement might look like. At key moments, they showed the promise of sustainable agriculture and the potential of sustainable agriculture initiatives to empower the poor to formulate their own responses to agrarian crisis that increase the autonomy of rural communities and challenge the control that powerful groups and institutions have over their lives. It is important to acknowledge these 'glimpses' of genuine alternatives, however fleeting they may have been, as they offer reason for hope. There were three levels at which the initiatives studied here showed promise in this regard: (1) in implementing effective strategies for *genuine participatory development*; (2) by *mobilising the grassroots* to build a knowledge-base for sustainable agriculture and to organise for change; (3) by *directly challenging hegemonic power and building the foundations of food sovereignty.*

The case studies showed that there are abundant opportunities for sustainable agriculture initiatives, even those led by 'outsiders', to facilitate rural people to compile their own knowledge about agrarian crisis and sustainability issues. KVM, for example, facilitated members of rural communities to build awareness of the impacts of the Green Revolution. This was particularly evident in their women's group, in which women documented changes in the health of their communities. This highlights the possibility of what might be termed 'grassroots research'—community-based initiatives to actively develop knowledge and document the need for alternatives. KVM's intervention was ultimately a good example of the 'participatory development' approach to facilitating grassroots knowledge production—complete with participatory development's democratic shortcomings. It empowered people with research skills to document changes in their communities and to use the knowledge produced in the process to intervene in policy debates. Furthermore, in the process of going from house-to-house to conduct their surveys, community participants built both awareness and social capital, which may in turn have assisted in the development of solutions. Yet, as critics have pointed out with respect to participatory development methods more broadly, the ultimate power lay with KVM as 'facilitators', as they dictated the ultimate parameters of the research. Rural participants certainly stand to gain from such an encounter, but it is hardly a meaningful challenge to hegemony, if both the process of inquiry and the alternative systems proposed continue to be framed by urban middle-class activists and development workers.

Revathi's attempt to facilitate grassroots, decentralised, farmer-led 'movements' deserves some attention as a potentially powerful instance of how participatory development methods could be applied in ways that are genuinely empowering. Revathi's vision in facilitating these 'movements' was not initially to politically empower the rural poor (though she did not foreclose that possibility). Rather, it aimed to elicit their participation in the development of sustainable agriculture, share knowledge, and collectively solve whatever other problems may arise, including marketing, collective bargaining, and campaigning. The structures Revathi facilitated would be decentralised to allow for maximum local control but also to create collective structures that would be beyond the capacity of the state to intervene in. She was an exceptionally astute organiser and learned from a number of failures regarding the need to avoid the traps of working within the development establishment. For example, after a number of negative experiences, she became disillusioned with 'community-based organisations', seeing them as vehicles for upper-caste dominance. She formulated a model of facilitating local collectives based on principles of economic empowerment and inter-dependency. Although it was still a 'work in progress' and faced a number of challenges, Revathi's model deserves further research attention. Through trial and error, she had devised means of working within the development establishment, while maintaining a view to facilitate something that would extend beyond it. The (still unrealised) end game was a self-sustaining movement for sustainable rural development and political empowerment of neglected sections of Tamil Nadu's rural society.

Sustainable agriculture initiatives also provide a medium for horizontal networking and knowledge sharing between farmers (Holt-Gimenez, 2006). KVM regularly hosts events in which natural farmers from across India run workshops for local farmers, theoretically providing horizontal networks through which farmers can exchange knowledge on natural farming techniques. It should be acknowledged, however, that these 'master farmers' tended to be educated larger land holders, making their homology with those with whom they network a little tenuous. Furthermore, although TOFarM developed much of its body of knowledge through engaging with practicing ecological farmers, Revathi's experience reveals some of the factors that may limit the potential of horizontal networking. The competitiveness she observed between sustainable farmers in the state restricted the potential for sharing innovations, while within communities, divisions of caste, class, religion, and gender undermined collective efforts.

At least one example identified within this research could unambiguously be identified as an instance of food sovereignty. BBA's seed-saving initiative was a collective effort on the part of small-holding farmers to solve problems relating to sustainability that were affecting their communities. Though Jardhari coordinated the effort, it remained collectivist in its approach, with farmers from throughout the region assisting in gathering traditional seeds and protecting local agro-biodiversity. In helping to redistribute local seeds, BBA's grassroots project brought tangible benefits to communities, some of which had lost access to many traditional varieties of seed during the 1970s. Locally managed seed banks are a versatile means of conserving agro-biodiversity and maintaining community control over genetic resources. Despite BBA's declining relevance over time, its initial seed collective was a clear instance of a counter-hegemonic food sovereignty movement, challenging the control of seed and chemical companies over villages' agriculture. It comes as one of several initiatives throughout India which encourage farmers to experiment with a range of crop varieties in their fields to determine which perform best in local conditions (Vijayalakshmi and Balasubramanian, 2004). Such initiatives encourage local people to actively participate in the development of more resilient agro-ecosystems.[40]

Within the Fabric of Hegemony

Despite these positive signs, all three case studies had links to established hegemonic powers. These links deflected their potential to represent subaltern perspectives and to build counter-hegemonic alternatives to the corporate food regime. Consequently, rather than building 'food sovereignty', sustainable agriculture initiatives in India have tended to be either ineffective, or overly accommodating to elite interests, or a means of integrating poor and struggling farmers into global food markets.

This finding is noteworthy, since, although these are only three case studies, they are, nonetheless, *telling* case studies. These were examples of sustainable

[40] It is not always possible, however, to develop these seed banks in such a participatory manner. In Punjab, for example, local varieties of seed have been almost completely replaced by hybrid varieties, so there is little scope for community participation in the collection and preservation of seed. KVM's seed bank was managed by one farmer, who had obtained the seeds from farmers in other states. There was a lack of ecological and social capital necessary for this to be a communal effort that could meaningfully contribute towards the food sovereignty of communities.

agriculture initiatives that one would expect to be, each for different reasons, *most likely* to have a counter-hegemonic trajectory. KVM's seemingly incorruptible charismatic leader had been strident in his critique of foreign donors and the state and adhered to a 'people's movement' model of organising. TOFarM, through its 'facilitative' model, had endeavoured to provide rural communities, particularly the rural poor, with the opportunity to shape their own movement for ecological agriculture, with its urban middle-class organiser taking a backseat to grassroots voices wherever possible. Finally, BBA was a grassroots movement, initiated and led by small farmers with strong rural participation. As Flyvbjerg (2006) outlines, one of the chief merits of case studies research is their capacity to falsify established theories. This is particularly true when they are 'critical case studies', selected as instances that one would expect to be *most likely* to confirm said theories. The fact that even these case studies, which one would expect to be among the *most likely* to be counter-hegemonic were found to have links to the hegemonic establishment which, in turn, undermined their counter-hegemonic potential, falsifies the claim that predominates in food sovereignty literature that sustainable agriculture initiatives are inherently counter-hegemonic. It makes it doubtful indeed that other such initiatives in India would be able to overcome those hegemonic influences and forge a movement that truly lives up to the emancipatory image presented in the 'food sovereignty' literature. Furthermore, the cases revealed some of the mechanisms by which that counter-hegemonic potential was derailed.

The case studies revealed how donor influence remains a key issue, even amongst relatively less-structured organisations. Revathi was, of the three case studies, the most dependent on external financial support from institutional donors and ultimately found their influence subversive to her agenda. Her experience suggested that while some donors were more rigid than others, attempts to implement sustained projects met with frustrations. In her sponsored project to facilitate community-based organisations, Revathi's donor was altogether unresponsive when she told them that their preferred model of organising was leading to elite capture and that micro-credit programmes were having unintended and highly detrimental side-effects. This unresponsiveness reflects the worrying reality that the flow of donor funds has the primary objective of serving a neoliberal development agenda. Despite the stated aims of development donors, the empowerment of the rural poor is often, at best, a secondary concern.

For other organisations, such as KVM, avoiding association with the corrupting influence of donor organisations has become a preoccupation. Nonetheless, the more KVM progressed in its objectives of promoting natural farming, the more farmers expected organisational support and the more KVM was gradually forced to adopt a more professional approach and accept donor funds, just like any other NGO. The case demonstrated how even sustainable agriculture initiatives with an ostensible objection to foreign donors' influence are ultimately led to accept donor funds, simply as a consequence of the resource requirements of promoting sustainable agriculture when farmers are not sufficiently enthusiastic to play a leading role. Both KVM and BBA also demonstrated that refusal to be totally dependent on donors can also lead to a greater reliance on other sources of support. These can be equally problematic, as is discussed below.

Relations with the state varied between the organisations studied here, yet were always seen to be problematic. In all three cases, the state, specifically the state-level governments, were not seen to be supportive of their overall objectives. In Tamil Nadu and Punjab, this was perhaps not surprising, given those state governments' commitments to the promotion of Green Revolution technologies. Revathi and KVM generally avoided direct engagement with the state, perceiving it to be unsupportive, though KVM did attempt to cultivate relations with members of all political parties, in the hope that this might build a longer term political consensus on the need for change. Revathi's engagement with the state was demonstrative of how even when one consciously avoids the state, it can retain a strong indirect influence. Revathi's conviction that the state government was opposed to her work and fear that it would interfere to cancel her receipt of foreign contributions led her to adopt a very cautious approach to organising. She adhered to a decentralised, localist approach to prevent the state from attacking her. While there may have been other strategic benefits of this (helping to build leadership capacity from below, for example), it had also precluded the possibility of a larger, more consolidated movement. Thus, even when it does not directly or deliberately interfere, the state continues to shape the terrain of civil society, through its coercive powers, real or perceived. Indeed, it is possible that the lack of mass movements for sustainable agriculture in India (of the kind that have proliferated in Latin America) may be partially attributable to sustainable agriculture's dependency on foreign donations, which the state has the right to block in India under the FCRA.

In Uttarakhand, BBA's disinterest in engaging with the state was unexpected, given the Government of Uttarakhand's active promotion of organic farming. The issue that BBA identified, however, was that the organic certification regime that the state government was promoting was only of interest to farmers selling their produce on the market—particularly exporters. This did not apply to BBA's mostly subsistence-oriented farmers. In many respects, this reflects the nature of the contemporary food regime. During the 1960s and 1970s, the state's core objective for rural development was to promote boosted production through the Green Revolution. Under the corporate food regime that has developed in more recent decades, however, the key objective is to promote integration within global food markets, to the ultimate benefit of global agri-business and finance capital. In this respect, organic farming is perhaps even better positioned than chemical farming, as it can be marketed to wealthier consumers in the global north (see Friedmann, 2005, who speaks of a 'corporate-environmental food regime'). This demonstrates how the state, by speaking a similar language to sustainable agriculture organisations, may attempt to guide their agenda towards work that is more amenable to hegemonic interests.

Given that the case studies presented here were 'hybrid organisations', combining features of social movements and NGOs (Hazenfeld and Gibron, 2006), the level of influence by the state and donor organisations was perhaps less than for the average rural development NGO. One of the most interesting findings of this research has been that this lack of financial dependency on the state and donors only increased these organisations' reliance on dispersed activist networks—both as a source of resources and moral support. This was found to be more problematic than one might expect, as these networks were dominated by the middle classes, most of whom had class interests that differed from those of the rural poor. These middle-class activists tended to guide movements towards confirming the assumptions of globally trending discourses within the sustainable farming movement (for example, the variety of extreme localism that informed KVM's 'natural farming' model), rather than using sustainable agriculture to meet the needs and aspirations of rural communities. Although these activist perspectives were by no means hostile towards the poor (as in cases of middle-class activism documented by Anjaria, 2009; Baviskar *et al*, 2006), over-reliance on these activist networks resulted in irrelevance at the grassroots. Middle-class activists' continued input and leadership of sustainable agriculture initiatives hampers the development of grassroots leadership and ultimately reproduces the subaltern position of the rural poor.

Finally, the influence of rural elites over sustainable agriculture projects continues to impede their counter-hegemonic potential. There are three distinct issues in this respect. First, sustainable agriculture discourses can, at times, descend into a kind of 'agrarian populism' that romanticises 'traditional village communities', overlooking the substantial injustices within communities along the lines of caste, gender, class, religion, and other factors. These romantic images may lead to sustainable agriculture interventions that overlook these imbalances of power and allow for rural elites to reap the benefits of their interventions—entrenching their local hegemonic power, rather than challenging it. Second, sustainable agriculture activists tend to position land holders as their main targets and neglect the perspectives of the landless. Given that capitalist land-holding farmers have been a lynchpin of the hegemonic establishment since the time of the Green Revolution, this does not appear to be upsetting the balance of hegemonic power—certainly not in the manner of the Latin American agroecology movements, which have a strong focus on land redistribution. Finally, in the instances when sustainable agriculture activists *do* seek to involve the poor, as in Revathi's case, this proves a highly challenging terrain when communities are fraught with internal divisions. Building a cohesive, community-wide response to agrarian crisis that challenges hegemonic relations is extremely challenging when elites may reprimand or even intimidate activists for engaging with rural subaltern groups.

As a result of this continued dominance of hegemonic influences in the formulation of sustainable agriculture projects, the subaltern remains speechless. That hegemonic influences persisted in these case studies, which one might expect to be more reflective of grassroots concerns than more conventional NGOs, suggests that hegemonic influence on mainstream sustainable rural development projects will be even more influenced by hegemonic forces. Certainly, other studies have found that larger NGOs working in this field in India are more oriented towards building global organic markets than building farmer autonomy, in the manner that the food sovereignty paradigm envisions. This has been found true even of NGOs, such as Vandana Shiva's Navdanya, which claim explicitly to be promoting food sovereignty (Trauger, 2015). In so doing, such NGOs effectively become part of the machinery of hegemony: they help construct *consent* for the corporate food regime, rather than build radical alternatives to it.

The implication of this finding is that we need to de-link sustainable agriculture from the counter-hegemonic struggle for food sovereignty. There is simply no necessary connection between the two—in fact, increasingly, sustainable agriculture initiatives appear to be a mechanism for holding the current hegemonic order in place. This finding, in turn, leads to two separate conclusions for two separate audiences. For those within the development establishment, it implies a need to reconsider their approach to sustainable farming interventions, given that their embeddedness within social relations that reproduce the status quo ultimately undermines their equity, effectiveness, and sustainability. For activists and scholars interested in the struggle for food sovereignty and against hegemony in rural India, it implies a need to consider other, less-obvious sites in which such struggles may be taking place.

Towards Inclusive and Sustainable Rural Development

For those working within the development establishment, the findings suggest a need to reorient approaches to sustainable rural development. Insofar as the development community shares a commitment to promoting rural development in an effective, equitable, and sustainable manner (sustainable on social, economic, and ecological fronts), there is a great need for more attention to the ways in which hegemonic influences permeate the institutional landscape of rural development.

For it must be acknowledged that these influences, often treated as a 'necessary evil', can actively *undermine* the objective of effective, equitable, and sustainable development interventions. Furthermore, recognising that the agents of hegemony come in the variety of guises listed above may help to ensure that attempts to avoid one source of influence does not result in one falling into the hands of another.

A major reason why sustainable agriculture initiatives in India are so prone to these sources of influence is their very method of formulating their objectives. Organisations, particularly those with urban middle-class leadership, enter communities with a pre-formulated agenda, rather than engaging with the rural poor, understanding their aspirations, and collaborating to achieve goals of mutual interest. This relates to what Jackson (1997) describes as the tension between sustainability and participatory development. The leading activists of these organisations, informed by donors, the wider activist community, and, in some cases, the state, have already decided on a development model that they believe is suitable—namely, ecological farming, free of the use of synthetic chemicals. Some organisations can be particularly rigid in their adherence to specific schools of sustainable farming (natural farming, permaculture, and so on). In such cases, rural participation is reduced to questions of how these models are implemented, and that, too, often comes under the influence of local elites. If rural communities were more involved in the formulation of what is to be done, the development models implemented might be more responsive to their immediate needs and hence more likely to be successful over a longer time frame (i.e., they will be more socially and economically sustainable). At present, most sustainable agriculture organisations are only participatory for instrumental reasons—they believe that encouraging local participation will make their projects have a more enduring impact. This should be turned on its head: participation and, more importantly, *empowerment* of the marginalised should be taken up as core objectives, with the principles of sustainability drawn on towards that end. This is challenging, however, when organisations depend on donors and middle-class supporters for patronage, to whom they must represent themselves as legitimate, authentic, virtuous, and so on. In time, this kind of self-promotion often overshadows and distorts the agenda of sustainable agriculture organisations and precludes possibilities for local agency.

Looking at examples of sustainable farming initiatives that have been more effective, equitable, and sustainable may provide indications of how these problems can be overcome. While these may not be 'counter-hegemonic' in the sense of being disruptive to power relations, they have nonetheless delivered far better outcomes—often because they negotiate the tension between sustainability and participation more carefully. The Centre for Sustainable Agriculture (CSA) in Telangana, for example, engaged directly with farmers in their region to understand which aspects of the chemical agricultural paradigm were not working for them and where change was most needed. They found that while there was no immediate interest in abandoning chemical fertilisers, there was a widespread disillusionment with pesticide usage, owing to rising pesticide costs and negative effects on health. CSA then promoted pesticide reduction through the use of non-pesticide management strategies, rather than adhering to a strict ideological position and demanding the removal of all chemical inputs (ala KVM). They continued with this strategy, despite criticism from their activist peers in the sustainable agriculture community for implementing 'half-baked' solutions. The results speak for themselves: approximately 350,000 farmers were using these techniques in the Telangana region as of 2007 (Ramanjaneyulu *et al*, 2009), and this number was reported to have risen to as many as a million farmers by 2015. Such strategies, whereby sustainable agriculture is promoted in a way that recognises the immediate needs and perspectives of rural communities rather than simply imposing models that are most congruent with activists' or donors' ideological visions (from 'natural farming' to micro-credit), are a far better approach to ensuring that low external input sustainable agriculture can move beyond the margins.

Such a change in approach needs to manifest in the everyday social interactions between rural development activists and rural community members. Taking the rural poor as one's chief collaborators, rather than relying on external activist networks for moral validation, will help to remove the boundaries that position development workers as 'outsiders'. When activists enter communities with the assumption that they know best and self-righteously assert the moral legitimacy of their own preferred farming models, this has the opposite effect: it *reinforces* activists' status as outsiders. This requires swallowing a very bitter pill—admitting that much of the activist and development literature on what is 'right' for rural communities often does not resonate with rural aspirations. Cultivating a willingness to genuinely listen is of the utmost importance, as the participatory development literature has repeatedly emphasised for decades now (e.g., Chambers, 1997). My research has found that activist networks play a major role in reinforcing a self-righteous disposition amongst activists that prevent authentic listening. Insofar as these activist networks will inevitably have some impact on activists' sense of identity and purpose, more attention needs to be directed towards the kind of culture and values that circulate within these networks. An effort needs to be made to shift the activist culture away from an emphasis on differences between agricultural philosophies and practices (itself a kind of 'identity politics') towards greater emphasis on the merits of listening, participation, and democratic engagement, particularly amongst the most marginalised.

Sustainable farming organisations also need to take more care in their relationship with rural elites. Before they begin their initiatives, they should gain a clearer sense of the heterogeneity of the communities in which they work and the distinct class interests of elites and subalterns. Some sustainable farming initiatives may prioritise working with elites in the hope that this may deliver quicker and easier results. They may feel that achieving ecological goals is the primary aim and the social means by which this is achieved are secondary. Development workers need to recognise that elites may leverage privileged positions within NGOs to exclude subordinate groups, particularly women and the landless. Prioritising working with more poor and marginal sections of communities from the outset of new projects is the obvious path to addressing this problem. Though this may be more challenging in the beginning, it will lead to more sustainable and far more equitable forms of change. Again, there are clear models of success in this regard, including amongst NGOs that work to empower those who are affected by multiple forms of social and economic disadvantage. The Deccan Development Society in Telangana has been highly successful in its sustainable farming endeavours and takes as its core constituency landless Dalit women. Furthermore, much of the progress of La Via Campesina in Latin America in promoting agroecology has occurred because of their engagements with issues affecting landless or near-landless rural labourers and tying their agroecology promotion to a long-term agenda for land redistribution (Borras and Franco, 2010). In the Indian context, however, many involved in the sustainable farming movement are reluctant to directly mobilise the most marginalised and are particularly averse to class-based mobilisation. This can be attributed to their adherence either to romantic images of the traditional village unit that gloss over intra-community tensions (Nanda, 2003) or Gandhian notions of 'trusteeship' that justify elite privilege by investing faith in elites' capacity for benevolence. Such ideas need to be interrogated more thoroughly at the grassroots level to avoid sustainable farming simply reproducing the status quo.

It must be openly acknowledged that working with the poor and marginalised is often very challenging: more so than engaging with elites. It is easy to say in theory that rural development workers should work to meet the aspirations of the rural poor, yet the fact remains that there is no simple method of determining what the rural poor 'really want'. Subaltern groups have long histories of exclusion and oppression and are, therefore, naturally sceptical of outsiders' attempts to facilitate change (Vincent, 2004). Additionally, the risks associated with publically stating the problems with the status quo provide a further incentive to remain silent (Masaki, 2010). Scholars of participatory development have therefore suggested that those attempting to facilitate change that genuinely empowers subaltern groups should begin with close, ethnographic studies of local situations (Masaki, 2004). Activists should attend to the multiple forms of what Scott (1985) terms 'everyday acts of resistance', specifically, the ways in which those who are too marginalised to explicitly challenge existing relations engage in subtle strategies to ensure their own survival and improve their social position. The sheer weight of agrarian crisis means that the rural poor, out of necessity, are already resisting their situation. Activists should look to these strategies that are already in practice, consider ways in which they can be expanded on, and engage in honest discussions about alternative strategies and the potentials and risks involved. Masaki (2004, 2010) argues that such strategies will be more effective than attempting to impose a development plan from above, as it activates and expands on practices that are not foreign to local people, being already immanent in local social relations.

The process of listening to and engaging with subaltern voices may ultimately take us beyond 'sustainable agriculture', narrowly defined. It requires a willingness to be far more negotiable in the development models promoted than has heretofore been the case. For example, it may require engaging with the non-agricultural livelihood strategies that have become an important tool for survival for the rural poor throughout much of India. As Ferguson (1994) identifies, development institutions have historically used representations of an isolated, unchanging, backward, intractably agricultural rural society as a means of depoliticising rural populations and justifying certain forms of intervention. Similarly, in representing India as an essentially agricultural nation and defending agriculture for its own sake, the organisations presented in this book neglected that the wider struggles of the rural poor for sustainable livelihoods involve the pursuit of both agricultural and non-agricultural livelihoods. A serious contribution towards rural struggles may require a wider conception of livelihoods and a more inter-sectoral perspective than is currently the case.

A disposition of openness towards multiple, locally-driven pathways to development, however, is highly challenging in the current institutional environment. Donors typically expect concisely formulated goals and quantifiable results. There is also no great enthusiasm within the development establishment for the focus on economic and political empowerment envisaged here. Nonetheless, this unfavourable institutional environment should not prevent scholars from putting forward possibilities for more radical departures from the status quo. The debate must continue to be pushed forward, lest rural development efforts remain short term, inequitable, and ineffective, as they all-too-often are at present.

Locating the Counter-Hegemonic in Rural India

Sustainable agriculture may well be a key component of counter-hegemonic struggles in some parts of the world. If reports are to be believed, in Latin America, it is a cornerstone of the food sovereignty movement, which actively opposes the corporate food regime. Yet, the case studies presented in this book and data emerging from other recent studies suggest that in India this is not the case.

Sustainable agriculture in India is so imbricated within hegemonic relations that, more often, it forms a link in the chain of the corporate food regime, rather than a radical alternative to it. This, in itself, is a fascinating finding, demonstrating the resilience of hegemony—the ways in which it reconfigures itself even in times of crisis and selectively incorporates elements of movements for what appear, at first glance, to be radical alternatives. Yet, it also implies that those interested in understanding counter-hegemonic movements in India, be they scholars or activists, should direct their attention elsewhere.

Indeed, to argue that sustainable agriculture is not, at present, a counter-hegemonic institution is by no means to suggest that there are no counter-hegemonic moments in contemporary rural India at all. It simply means that, if we are to understand how power relations in rural India are being contested, particularly those that are relevant to understanding hegemony, we must look to places less obvious than those that food-regime analysis identifies. Indeed, rather than looking for a kind of local variety of food sovereignty discourses that reject and subvert the global power of transnational agri-business, we might instead look to more localised challenges to long-established power relations, which are perhaps more unsettling to the hegemonic order that has been in place for decades.

The rural poor in India are far from passive. The Maoist movement, though in major decline at present, had demonstrated the deadly opposition that was possible from the rural poor to the systemic violence of neoliberalism in the countryside. Yet, there are also far more quotidian forms of resistance that nonetheless, when aggregated, constitute a major threat to established power. Indeed, contemporary India is abuzz with change, and the rural poor are a major part of that. Jeffrey (2017) describes India as passing through a 'social revolution', fuelled by increased access to education, new communication technologies, and new political opportunities. While the former two have greatly increased all people's capacity to mobilise, the latter—new political opportunities—have been particularly important for the rural poor.

Three developments that have occurred roughly in the past two decades have been incredibly important in giving the rural poor political opportunities to challenge the existing structure of power. The first relates to representation. What is referred to as 'democratic decentralisation'—the devolution of democratic decision-making and service provision to local levels, which has increased in India since the mid-1990s—is opening up political spaces in rural society of which the poor and under-represented are taking advantage. Representation on local panchayat boards gives the rural subaltern a say in ways that have not been possible in the past and, ironically, as the traditional elites migrate away from the countryside, the proportionate representation of the rural subaltern on these boards has only increased (Tanabe, 2007). The second development is a series of new welfare schemes, which the rural poor have, with increasing confidence, accessed and directed. Most notable in this respect is the National Rural Employment Guarantee Act (2006) (NREGA), which guarantees rural families up to hundred hours of waged work per year at the official minimum wage, often to work on rural development projects. Rural communities have been active, not only in ensuring that they access their entitlements under NREGA but also in directing NREGA projects towards work that will have lasting local benefits. Having access to this waged labour provides a kind of social security net for labouring rural families, preventing them from having to take on more exploitative work. The third development relates to accountability. New legislation, such as the Right to Information Act (2005) (RTI), has given rural citizens legal mechanisms for demanding information regarding their entitlements from governments and also allowing the discernment of where funds have been syphoned off by corrupt officials. Grassroots RTI movements have been influential in recent years in both the economic and political empowerment of the rural poor.

These kinds of political activity may not present as resistance to the corporate food regime. They may not appear as epic or inspiring as struggles for 'food sovereignty', and they may not appeal to activists' or intellectuals' political or aesthetic standards regarding what counter-hegemonic movements should look like. Yet, these micro-political acts are reconfiguring the power relations in which food is produced at the ground level. This is the terrain on which hegemony is being contested in contemporary rural India.

Bibliography

Achuthan, N. S. 2009. 'Four Years Beyond Tsunami: Contours of a Roadmap for a Coordinated "Multi-hazard (Including Tsunami) Risk Management Action Plan for Tsunami-affected Villages in Tamil Nadu": Overview of Ongoing/Projected Initiatives.' *Disaster Prevention and Management*, 18 (3): 249–269.

Agrawal, R. 2013. 'Hydropower Projects in Uttarakhand: Displacing People and Destroying Lives.' *Economic and Political Weekly* 48 (29): 14–16.

AID. 2012. *Association for India's Development*. Accessed 10 April 2012. Available at http://www.aidindia.org/main/.

Alexander, K. C. 1975a. 'Genesis of Agrarian Tensions in Thanjavur: Findings of a Research Study.' *Economic and Political Weekly* 10 (49): 1881–1886.

————. 1975b. 'Characteristics of the Agrarian Social Structure of Tamil Nadu.' *Economic and Political Weekly* 10 (16): 664–672.

Altieri, M. A. 1984. 'Towards a Grassroots Approach to Rural Development in the Third World.' *Agriculture and Human Values* 1 (4): 45–48.

————. 2002. 'Agroecological Principles for Sustainable Agriculture.' In *Agroecological Innovations: Increasing Food Production with Participatory Development*, edited by N. Uphoff, 41–46. London: Earthscan.

————. 2004. 'Linking Ecologists and Traditional Farmers in the Search for Sustainable Agriculture.' *Frontiers in Ecology and the Environment* 2(1): 35–42.

Altieri, M. A. and V. M. Toledo. 2011. 'The Agroecological Revolution in Latin America: Rescuing Nature, Ensuring Food Sovereignty and Empowering Peasants.' *Journal of Peasant Studies* 38 (3): 587–612.

Alvares, C. 2009. *The Organic Farming Sourcebook* (3rd Edition). Goa: Other India Press.

Anjaria, J. S. 2009. 'Guardians of the Bourgeois City: Citizenship, Public Space, and Middle-class Activism in Mumbai.' *City & Community* 8 (4): 391–406.

Appadurai, A. 2002. 'Deep Democracy: Urban Governmentality and the Horizon of Politics.' *Public Culture* 14 (1): 21–47.

Arnold, D. 1982. 'Rebellious Hillmen: The Gudem-Rampa Risings 1839–1924.' In *Subaltern Studies 1*, edited by R. Guha, 88–142. Oxford: Oxford University Press.

Arya, A. S., G.S Mandal, and E. V. Muley. 2006. 'Some Aspects of Tsunami Impact and Recovery in India.' *Disaster Prevention and Management* 15 (1): 51–66.

Baker, C. J. 1984. 'Frogs and Farmers: The Green Revolution in India and its Murky Past.' In *Understanding Green Revolutions: Agrarian Change and Development Planning in South Asia*, edited by T. P. Bayliss-Smith and S. Wanmali, 37–52. Cambridge: Cambridge University Press.

Balasubramanian, M. and P. J. D. Kumar. 2014. 'Climate Change, Uttarakhand, and the World Bank's Message.' *Economic and Political Weekly* 49 (1): 65–68.

Bardhan, P. 1970. '"Green Revolution" and Agricultural Labourers.' *Economic and Political Weekly* 5 (29–31): 1239–1246.

Bartley, T. 2007. 'How Foundations Shape Social Movements: The Construction of an Organizational Field and the Rise of Forest Certification.' *Social Problems* 54 (3): 229–255.

Barvale, R. 2009. Untitled. Press statement released 29 October 2009. Accessed 10 January 2010. Available at ftp://mahyco.com/mahycoweb/html/btBrinjalNews1.html.

Bates, R. H. 2005. *Markets and States in Tropical Africa: The Political Basis of Agricultural Policies* (2nd edition). Berkeley, California: University of California Press.

Batliwala, S. 2002. 'Grassroots Movements as Transnational Actors: Implications for Global Civil Society. *Voluntas* 13 (4): 393–409.

Baumann, P. 1998. 'The Persistence of Populism in Indian Forest Policy.' *Journal of Peasant Studies* 25 (4): 96–123.

Baviskar, A. 2004. *In the Belly of the River: Tribal Conflicts over Development in the Narmada Valley* (2nd edition). Oxford: Oxford University Press.

————. 2005. 'Red in Tooth and Claw? Looking for Class in Struggles over Nature.' In *Social Movements in India: Poverty, Power and Politics*, edited by R. Ray and M. F. Katzenstein, 161–178. New York: Rowman and Littlefield Publishers.

————. 2011. 'Cows, Cars and Cycle-rickshaws: Bourgeois Environmentalists and the Battle for Delhi's Streets.' In *Elite and Everyman: The Cultural Politics of India's Middle Classes*, edited by A. Baviskar and R. Ray, 391–418. New Delhi: Routledge.

Baviskar, A., S. Sinha, and K. Phillip. 2006. 'Rethinking Indian Environmentalism: Industrial Pollution in Delhi and Fisheries in Kerala.' In *Forging Environmentalism: Justice, Livelihood and Contested Environments*, edited by J. Bauer, 189–256. London: M. E. Sharpe.

Bernstein, H. 1996. 'Agrarian Questions Then and Now.' *Journal of Peasant Studies* 24 (1): 22–59.

————. 2009. 'Agrarian Questions from Transition to Globalisation.' In *Peasants and Globalization: Political Economy, Rural Transformation and the Agrarian Question*, edited by A. H. Akram-Lodhi and C. Kay, 239–261. London: Routledge.

Bhalla, G. S. and G. K. Chadha. 1983. *Green Revolution and the Small Peasant: A Study of Income and Distribution among Punjab Cultivators*. New Delhi: Concept Publishing Company.

Bhalla, S. 1976. 'New Relations of Production in Haryana Agriculture.' *Economic and Political Weekly* 11 (13).

Bharti, V. 2012. 'Indebtedness and Suicides: Field Notes on Agricultural Labourers of Punjab.' *Economic and Political Weekly* 46 (14): 35–40.

Bob, C. 2002. 'Merchants of Morality.' *Foreign Policy* 129: 36–45.

Bora, R. S. 1996. *Himalayan Migration: A Study of the Hill Region of Uttar Pradesh*. Delhi: Sage Publications.

Borras, S. M. and J. C. Franco. 2010. 'Food Sovereignty and Redistributive Land Policies: Exploring Linkages, Identifying Challenges.' In *Food Sovereignty: Reconnecting Food, Nature and Community*, edited by H. Wittman, A. Desmarais, and N. Wiebe, 106–119. Halifax: Fernwood Publishing.

Brass, T. 1997. 'The Agrarian Myth, the "New" Populism and the "New" Right.' *Journal of Peasant Studies* 24 (4): 201–245.

Brown, L. 1968. 'The Agricultural Revolution in Asia.' *Foreign Affairs* 46 (4): 688–698.

————. 1970. *Seeds of Change: The Green Revolution and Development in the 1970s*. New York: Praeger Publications.

Brown, T. 2013. 'Agrarian Crisis in Punjab and "Natural Farming" as a Response.' *South Asia: Journal of South Asian Studies* 36 (2): 229–242.

————. 2014a. 'Chipko Legacies: Sustaining an Ecological Ethic in the Context of Agrarian Change.' *Asian Studies Review* 38 (4): 639–657.

————. 2014b. 'Negotiating the NGO-social Movement Dichotomy: Evidence from Punjab, India.' *Voluntas* 25 (1): 46–66.

————. 2016. 'Sustainability as Empty Signifier: Its Rise, Fall, and Radical Potential.' *Antipode* 48 (1): 115–133.

————. 2017. 'Beyond the Participatory Development Impasse: Pathways Forward for Sustainable Rural Development.' *Social Transformations: Journal of the Global South* 5 (1): 3–30.

Bunch, R. and G. Lopez. 2005. 'Soil Recuperation in Central America: How Innovation was Sustained after Project Intervention.' In *The Earthscan Reader in Sustainable Agriculture*, edited by J. Pretty, 341–351. London: Earthscan.

Buttel, F. H. 1993. 'The Sociology of Agricultural Sustainability: Some Observations on the Future of Sustainable Agriculture.' *Agriculture, Ecosystems and Environment* 46 (1–4): 175–186.

Buttigieg, J. 2005. 'The Contemporary Discourse on Civil Society: a Gramscian Critique.' *Boundary 2* 32 (1): 33–52.

Byres, T. J. 1972. 'The Dialectic of India's Green Revolution.' *South Asian Review* 5 (2): 99–115.

————. 1981. 'The New Technology, Class Formation and Class Action in the Indian Countryside.' *Journal of Peasant Studies* 8 (4): 405–454.

————. 1991. 'The Agrarian Question and Differing Forms of Capitalist Agrarian Transition: An Essay with Reference to Asia.' In *Rural Transformation in Asia*, edited by J. Breman and S. Mundle. Delhi: Oxford University Press.

Chakravarty, S. 2008. *Red Sun: Travels in Naxalite Country*. New Delhi: Penguin.

Chambers, R. 1997. *Whose Reality Counts? Putting the First Last*. London: ITDG Publishing.

Chand, R. 1999. 'Emerging Crisis in Punjab Agriculture: Severity and Options for Future. *Economic and Political Weekly* 34 (13): A2–A10.

————. 2009. 'Capital Formation in Indian Agriculture: National and State Level Analysis.' In *Agrarian Crisis in India*, edited by D. N. Reddy and S. Mishra. Oxford: Oxford University Press.

Chandhoke, N. 2009. 'Putting Civil Society in its Place.' *Economic and Political Weekly* 44 (7): 12–16.

Chatterjee, P. 1982. 'Agrarian Relations and Communalism in Bengal, 1926–1935.' In *Subaltern Studies 1*, edited by R. Guha, 9–38. Oxford: Oxford University Press.

————. 1998. 'Development Planning and the Indian State.' In *The State, Development Planning and Liberalisation in India*, edited by T. J. Byres, 82–103. Oxford: Oxford University Press.

————. 2004. *Politics of the Governed: Reflections on Popular Politics in Most of the World*. New York: Columbia University Press.

———. 2011. *Lineages of Political Society: Studies in Postcolonial Democracy*. New York: Columbia University Press.

Cleaver, H. M. 1972. 'The Contradictions of the Green Revolution.' *The American Economic Review* 62 (1–2): 177–186.

Cohen, J. L. and A. Arato. 1994. *Civil Society and Political Theory*. Cambridge, Massachusetts: MIT Press.

Conquest, R. 1986. *The Harvest of Sorrow: Soviet Collectivisation and the Terror-Famine*. London: Hutchinson.

Conway, G. 1985. 'Agroecosystem analysis.' *Agricultural Administration* 20 (1): 31–55.

Conway, G. and E. B. Barbier. 1990. *After the Green Revolution: Sustainable Agriculture for Development*. London: Earthscan.

Cooke, B. 2004. 'Rules of Thumb for Participatory Change Agents.' In *Participation: from Tyranny to Transformation?*, edited by S. Hickey and G. Mohan, 42–55. London: Zed Books.

Cooke, B. and U. Kothari. 2001. 'The Case for Participation as Tyranny.' In *Participation: The New Tyranny?*, edited by B. Cooke and U. Kothari. London: Zed Books.

Cooley, A. and J. Ron. 2002. 'The NGO Scramble: Organizational Insecurity and the Political Economy of Transnational Action.' *International Security* 27 (1): 5–39.

Corbridge, S., J. Harriss, and C. Jeffrey. 2013. *India Today: Economy, Politics and Society*. Cambridge: Polity.

Cunich, S. 2012. *Growing Change* (documentary). Sydney: Simon Cunich.

Dabholkar, S. 2001. *Plenty for All: Prayog Pariwar Methodology* (2nd Edition). Pune: Mehta Publishing House.

Dagnino, E. 2008. 'Challenges to Participation, Citizenship and Democracy: Perverse Confluence and Displacement of Meanings.' In *Can NGOs Make a Difference? The Challenge of Development Alternatives*, edited by A. Bebbington, S. Hickey, and D. C. Mitlin, 55–70. London: Zed Books.

Darnhofer, I., W. Schneeberger, and B. Freyer. 2003. 'Converting or Not Converting to Organic Farming in Austria: Farmer Types and Their Rationale.' *Agriculture and Human Values* 22 (1): 39–52.

de Alcantara, C. H. 1974. 'The "Green Revolution" as History: The Mexican Experience.' *Development and Change* 5 (2): 25–44.

de León, A. G. 1995. 'Galloping into the Future.' In *First World, ha ha ha!*, edited by E. Katzenberger, 211. San Francisco: City Lights Books.

de Wit, J. and E. Berner. 2009. 'Progressive Patronage? Municipalities, NGOs, CBOs and the Limits to Slum Dwellers' Empowerment.' *Development and Change* 40 (5): 927–947.

Denscombe, M. 2007. *The Good Research Guide for Small Scale Social Research Projects*. Maidenhead: McGraw-Hill Education

Department of Agriculture and Co-operation. 2010. *Annual Report: 2009–2010*. Retrieved on 3 July 2012. Available at http://agricoop.nic.in/AnnualRepor2009-10/AR.pdf.

Desai, D. K. 1969. 'Intensive Agricultural District Programme: Analysis of Results.' *Economic and Political Weekly* 4 (26): A83–A90.

Deshpande, R. 2004. 'Social Movements in Crisis?' In *Indian Democracy: Meanings and Practices*, edited by R. Vora and S. Palshikar, 379–409. London: Sage Publications.

Deshpande, S. 2003. *Contemporary India: A Sociological View*. New Delhi: Viking.

Desmarais, A. 2002. 'The Vía Campesina: Consolidating an International Peasant and Farm Movement.' *Journal of Peasant Studies* 29 (2): 91–124.

———. 2003. 'La Vía Campesina: Peasant Women on the Frontiers of Food Sovereignty.' *Canadian Woman Studies* 23 (1): 140–145.

———. 2007. *La Via Campesina: Globalisation and the Power of Peasants.* London: Pluto Press.

Devereux, S. and J. Hoddinott. 1993. 'Issues of Data Collection.' In *Fieldwork in Developing Countries*, edited by S. Devereux and J. Hoddinott. Boulder, Colorado: Lynne Rienner Publishers.

Dill, B. 2009. 'The Paradoxes of Community-based Participation in Dar es Salaam.' *Development and Change* 40 (4): 717–743.

Djurfeldt, G., V. Athreya, N. Jayakumar, S. Lindberg, A. Rajagopal, and R. Vidyasagar. 2008. 'Agrarian Change and Social Mobility in Tamil Nadu.' *Economic and Political Weekly* 43 (45): 50–61.

Dogra, B. 2000. 'Linking Environment Protection to People's Livelihood.' *Frontier* 33 (19): 6.

Dwyer, A. 2011. *Broke but Unbroken.* Halifax: Fernwood Publishing.

Edelman, M. 2009. 'Synergies and Tensions between Rural Social Movements and Professional Researchers.' *Journal of Peasant Studies* 39 (1): 245–265.

Eisinger, P. K. 1973. 'The Conditions of Protest Behavior in American Cities.' *The American Political Science Review* 67 (1): 11–28.

Ekins, P. 1992. *A New World Order: Grassroots Movements for Global Change.* London: Routledge.

Escobar, A. 1992. 'Reflections on Development: Grassroots Approaches and Alternative Politics in the Third World.' *Futures* 24 (5): 411–436.

———. 1998. 'Whose Knowledge, Whose Nature? Biodiversity, Conservation and the Political Ecology of Social Movements.' *Journal of Political Ecology* 5 (1): 53–82.

Fairbairn, M. 2012. 'Framing Transformation: The Coutner-hegemonic Potential of Food Sovereignty in the US Context.' *Agriculture and Human Values* 29 (2): 217–230.

Fairweather, J. R. 1999. 'Understanding How Farmers Choose between Organic and Conventional Production: Results from New Zealand and Policy Implications.' *Agriculture and Human Values* 16 (1): 51–63.

Farooquee, N. A. and R. K. Maikhuri. 2007. 'Role of the State on Forests: Case of Uttarakhand.' *Economic and Political Weekly* 42 (35): 3537–3540.

Farrington, J. and A. Bebbington. 1993. *Reluctant Partners? Non-Government Organizations, the State and Sustainable Agricultural Development.* London: Routledge.

Farrington, J. and D. J. Lewis. 1993. *Non-Governmental Organizations and the State in Asia.* London: Routledge.

Ferguson, J. 1994. *The Anti-Politics Machine: 'Development,' Depoliticization, and Bureaucratic Power in Lesotho.* Minneapolis: University of Minnesota Press.

Fernandes, L. 2006. *India's New Middle Class.* Minneapolis: University of Minnesota Press.

Fernandes, L. and P. Heller. 2008. 'Hegemonic Aspirations: New Middle Class Politics and India's Democracy in Comparative Perspective.' In *Whatever Happened to Class? Reflections from South Asia*, edited by R. J. Herring and R. Agarwala, 146–165. London: Routledge.

B. Flyvbjerg. 2006. 'Five Misunderstandings about Case-study Research.' *Qualitative Inquiry* 12 (2): 219–245.

Fontana, B. 2006. 'Liberty and Domination: Civil Society in Gramsci.' *Boundary 2* 33 (2): 51–74.

Frankel, F. R. 1971. *India's Green Revolution: Economic Gains and Political Costs*. Princeton: Princeton University Press.

Friedmann, H. 1978. 'World Market, State, and Family Farm: Social Bases of Household Production in the Era of Wage Labor.' *Comparative Studies in Society and History* 10 (4): 545–586.

————. 1993. 'The Political Economy of Food: A Global Crisis.' *New Left Review* 197: 29–57.

————. 2005. 'From Colonialism to Green Capitalism: Social Movements and Emergence of Food Regimes.' *Research in Rural Sociology and Development* 11: 227–264.

Fukuoka, M. 2009. *One Straw Revolution*. Translated by L. Korn. New York: NYRB Classics.

Funder, M. 2010. 'The Social Shaping of Participatory Spaces: Evidence from Community Development in Southern Thailand.' *Journal of Development Studies* 46 (10): 1708–1728.

Galab, S., E. Revathi, and P. P. Reddy. 2009. 'Farmers' Suicides and Unfolding Agrarian Crisis in Andhra Pradesh.' In *Agrarian Crisis in India*, edited by D. N. Reddy and S. Mishra, 164–198. New Delhi: Oxford Universiy Press.

Galvin, S. S. 2014. 'Organic Designs and Agrarian Practice in Uttarakhand, India.' *Culture, Agriculture, Food and Environment* 36 (2): 118–128.

Gandhi, M. K. 1963. *Village Swaraj*. Edited by H. M. Vyas. Ahmedabad: Navajivan Publishing House.

Gerber, J. and S. Veuthey. 2010. 'Plantations, Resistance and the Greening of the Agrarian Question in Coastal Ecuador.' *Journal of Agrarian Change* 10 (4): 455–481.

Ghosh, S. 2009. 'NGOs as Political Institutions.' *Journal of Asian and African Studies* 44 (5): 475–495.

Gilbert, E. G., D. W. Norman, and F. E. Winch. 1980. *Farming Systems Research: A Critical Appraisal*. East Lansing: Michigan State University.

Gill, S. S. 2000. 'Agrarian Change and Struggle for Rural Labour in Punjab.' *International Journal of Punjab Studies* 7 (1): 73–91.

————. 2004. 'Punjab: Farmers' Movement: Continuity and Change.' *Economic and Political Weekly* 39 (27): 2964–2966.

————. 2005. 'Economic Distress and Farmer Suicides in Rural Punjab.' *Journal of Punjab Studies* 12 (2): 219–237.

Gill, S. S. and K. C. Singhal. 1984. 'Farmers Agitation: Response to Development Crisis of Agriculture.' *Economic and Political Weekly* 19 (40): 1728–1732.

Gopikrishnan, S. R. 2012. 'Soil Health and Support Systems: Contradictions and Missing Links.' *Economic and Political Weekly* 47 (29): 24–26.

Gramsci, A. 1971. *Selections from the Prison Notebooks of Antonio Gramsci*. Edited and translated by Q. Hoare and G. N. Smith. London: Lawrence and Wishart.

Gray, R., J. Bebbington, and D. Collison. 2006. 'NGOs, Civil Society and Accountability: Making the People Accountable to Capital.' *Accounting, Auditing and Accountability Journal* 19 (3): 319–348.

Guha, R. 2000. *The Unquiet Woods: Ecological Change and Peasant Resistance in the Himalaya* (2nd Edition). Berkeley: University of California Press.

———. 2007. *India After Gandhi: The History of the World's Largest Democracy*. London: Picador.

Guha, R., and J. Martinez-Alier. 1997. *Varieties of Environmentalism: North and South*. London: Earthscan Publications.

Guha, R. 1982. 'On Some Aspects of the Historiography of Colonial India.' In *Subaltern Studies 1*, edited by R. Guha, 1–7. Oxford: Oxford University Press.

———. 1998. *Dominance without Hegemony: History and Power in Colonial India*. Delhi: Oxford University Press.

Gupta, A. 1998. *Postcolonial Developments: Agriculture in the Making of Modern India*. Durham: Duke University Press.

Gupta, A. K. 2013. 'Tapping the Entrepreneurial Potential of Grassroots Innovation.' *Stanford Social Innovation Review* 11 (3): 18–20.

Handa, O. C. 2002. *History of Uttaranchal*. New Delhi: Indus Publishing Company.

Harriss, J. 2007. 'Antinomies of Empowerment: Observations on Civil Society, Politics and Urban Governance in India.' *Economic and Political Weekly* 45: 2716–2724.

———. 2008. 'Middle Class Activism and the Politics of the Informal Working Class: A Perspective on Class Relations and Civil Society in Indian Cities.' In *Whatever Happened to Class? Reflections from South Asia*, edited by R. J. Herring and R. Agarwala, 109–126. London: Routledge.

Harriss, J., J. Jeyaranjan, and K. Nagaraj. 2010. 'Land, Labour and Caste Politics in Rural Tamil Nadu in the 20th Century: Iruvelpattu (1916–2008).' *Economic and Political Weekly* 45 (13): 47–61.

Harriss-White, B. and S. Janakarajan. 1997. 'From Green Revolution to Rural Industrial Revolution in South India.' *Economic and Political Weekly* 32 (25): 1469–1477.

Hasenfeld, Y. and B. Gidron. 2005. 'Understanding Multi-purpose Hybrid Voluntary Organisations: The Contributions of Theories on Civil Society, Social Movements and Non-profit Organisations.' *Journal of Civil Society* 1 (2): 97–112.

Headley, D. and S. Fan. 2008. 'Anatomy of a Crisis: The Causes and Consequences of Surging Food Prices.' *Agricultural Economics* 39 (Supplement): 375–391.

Herring, R. J. 1981. 'Embedded Production Relations and the Rationality of Tenant Quiescence in Tenure Reforms.' *Journal of Peasant Studies* 8 (2): 131–172.

———. 1983. *Land to the Tiller: The Political Economy of Agrarian Reform in South Asia*. New Haven: Yale University Press.

———. 2008. 'Why did "Operation Cremate Monsanto" Fail? Science and Class in India's Great Terminator-technology Hoax.' In *Whatever Happened to Class? Reflections from South Asia*, edited by R. J. Herring and R. Agarwala, 127–145. London: Routledge.

Hobsbawm, E. 1973. 'Peasants and Politics.' *Journal of Peasant Studies* 1 (1): 3–22.

Howson, R. 2006. *Challenging Hegemonic Masculinity*. London: Routledge.

Holt-Giménez, E. 2006. *Campesino a Campesino: Voices from Latin America's Farmer to Farmer Movement for Sustainable Agriculture*. Oakley, California: Food First Books.

Holt-Giménez, E. and M. A. Altieri. 2013. 'Agroecology, Food Sovereignty and the New Green Revolution.' *Agroecology and Sustainable Food Systems* 37 (1): 90–102.

Isakson, S. R. 2009. 'No Hay Ganancia en la Milpa: The Agrarian Question, Food Sovereignty, and the On-farm Conervation of Agrobiodiversity in the Guatamalan Highlands.' *Journal of Peasant Studies* 36 (4): 725–759.

Ishizaka, S. 2014. 'Re-evaluating the Chipko (Forest Protection) Movement in India.' *The South Asianist* 2 (1): 9–27.

Jackson, C. 1997. 'Sustainable Development at the Sharp End.' *Development in Practice* 7 (3): 237–247.

Jakimow, T. 2010. 'Negotiating the Boundaries of Voluntarism.' *Voluntas* 21 (4): 546–568.

Jalali, R. 2008. 'International Funding of NGOs in India: Bringing the State Back in.' *Voluntas* 19 (2): 161–88.

James, G. A. 2013. *Ecology is Permanent Economy: The Activism and Environmental Philosophy of Sunderlal Bahuguna*. Albany, New York: SUNY Press.

Jardhari, V. n.d. *Baranaja*. Raigad, Maharashtra: Academy of Development Science.

Jeffrey, C. 2017. *Modern India: A Very Short Introduction*. Oxford: Oxford University Press.

Jeromi, P. D. 2007. 'Farmers' Indebtedness and Suicides: Impact of Agricultural Trade Liberalisation in Kerala.' *Economic and Political Weekly* 42 (31): 3241–3247.

Jodhka, S. S. 1997. '"Crisis" of the 1980s and Changing Agenda of "Punjab Studies": A Survey of Some Recent Research.' *Economic and Political Weekly* 32 (6): 273–279.

————. 2006. 'Beyond "Crises": Rethinking Contemporary Punjab Agriculture.' *Economic and Political Weekly* 41 (16): 1530–1537.

————. 2012. 'Agrarian Changes in the Times of (Neo-liberal) Crises: Revisiting Attached Labour in Haryana.' *Economic and Political Weekly* 47 (26–27): 5–13.

————. 2015. 'Punjab at the Crossroads.' *Economic and Political Weekly* 50 (42): 17–19.

Kamat, S. 2002. *Development Hegemony: NGOs and the State in India*. New Delhi: Oxford University Press.

Kantowsky, D. 1980. *Sarvodaya: The Other Development*. New Delhi: Vikas Publishing House.

Kapur, D. 2011. 'The Shift to Cash Transfers: Running Better but on the Wrong Road?' *Economic and Political Weekly* 46 (21): 80–85.

Kautsky, K. 1988. *The Agrarian Question*. London: Zwan Publications.

Kaviraj, S. 1986. 'Indira Gandhi and Indian Politics.' *Economic and Political Weekly* 21 (38–39): 1697–1708.

Khosla, R. 2006. 'A Participatory Organic Guarantee Scheme for India.' Retrieved from http://www.ifoam.org/about_ifoam/standards/pgs_projects/pgs_downloads/PGSforIndia_FinalReport.pdf.

Klenk, R. M. 2010. *Educating Activists: Development and Gender in the Making of Modern Gandhians*. New York: Lexington Books.

Kochhar, N., G. S. Gill, N. Tuli, V. Dadwal, and V. Balaram. 2007. 'Chemical Quality of Ground Water in Relation to Incidence of Cancer in Parts of SW Punjab, India.' *Asian Journal of Water, Environment and Pollution* 4 (2): 107–112.

Kohli, A. 1987. *The State and Poverty in India: The Politics of Reform*. Cambridge: Cambridge University Press.

Kohli, K. 2004. 'Mine, not Yours.' *The Hindu*, 28 March.

Krishna, S. 2002. 'Economic Profile of Uttaranchal.' *Economic and Political Weekly* 37 (19): 1843–1849.

Kulkarni, H. and M. Shah. 2013. 'Punjab Water Syndrome: Diagnostics and Prescriptions.' *Economic and Political Weekly* 48 (52): 64–73.

Kumar, A. 2007. 'The 2007 Punjab Election: Exploring the Verdict.' *Journal of Punjab Studies* 14 (1): 267–280.

———. 2011. *The Making of a Small State: Populist Social Mobilisation and the Hindi Press.* New Delhi: Orient Black Swan.

Kuruganti, K. 2010. 'Un-scientific Dismissal.' *Economic* and *Political Weekly* 45 (25).

Ladha, J. K., *et al.* 2003. 'How Extensive are Yield Declines in Long-term R–wheat Experiments in Asia?' *Field Crops Research* 81 (2–3): 159–180.

Lears, T. J. 1985. 'The Concept of Cultural Hegemony: Problems and Possibilities.' *The American Historical Review* 90 (3): 567–593.

Lerche, J. 2011. 'Review Essay: Agrarian Crisis and Agrarian Questions in India.' *Journal of Agrarian Change* 11 (1): 104–118.

Levien, M. 2011. 'Special Economic Zones and Accumulation by Dispossession in India.' *Journal of Agrarian Change* 11 (4): 454–483.

Lewis, J. W. 1974. *Peasant Rebellion and Communist Revolution in Asia.* Stanford, California: Stanford University Press.

Ludden, D. 1999. *An Agrarian History of South Asia.* Cambridge: Cambridge University Press.

Luthra, S. 2003. 'Educating Entrepreneurs, Organizing for Social Justice: NGO Development Strategies in New Delhi Bastis.' In *Regional Modernities: The Cultural Politics of Development in India*, edited by A. Agrawal and K. Sivaramakrishnan, 237–261. Stanford, California: Stanford University Press.

Martin, J. 1997. 'Hegemony and the Crisis of Legitimacy in Gramsci.' *History of the Human Sciences* 10 (1): 37–56.

Martínez-Torres, M. E. and P. Rosset. 2010. 'La Vía Campesina: The Birth and Evolution of a Transnational Social Movement.' *Journal of Peasant Studies* 39 (1): 149–175.

Marx, K. 2005. *The Eighteenth Brumaire of Louis Bonaparte.* New York: Mondial.

Masaki, K. 2004. 'The "Transformative" Unfolding of "Tyrannical" Participation: The *Corvée* Tradition and Ongoing Local Politics in Western Nepal.' In *Participation: From Tyranny to Transformation?*, edited by S. Hickey and G. Mohan, 125–139. London: Zed Books.

———. 2010. 'Rectifying the Anti-politics of Citizen Participation: Insights from the Internal Politics of a Aubaltern Community in Nepal.' *Journal of Development Studies* 46 (7): 1196–1215.

Mawdsley, E. 1998. 'After Chipko: From Environment to Region in Uttaranchal.' *Journal of Peasant Studies* 25 (4): 36–54.

McAdam, D. 1999. *Political Process and the Development of Black Insurgency, 1930–1970* (2nd Edition). Chicago: The University of Chicago Press.

McCarthy, J. D. and M. N. Zald. 1977. 'Resource Mobilization and Social Movements: A Partial Theory.' *American Journal of Sociology* 82 (6): 1212–1241.

McHattie, B. 2000. *Threats Posed by Globalisation and Responses by Rural Social Movements: Case of the Karnataka Rajya Raita Sangha in South India.* M.A. thesis, The Faculty of Graduate Studies, University of Guelph.

McLaverty, P. 2002. 'Civil Society and Democracy.' *Contemporary Politics* 8 (4): 303–318.

McMichael, P. 2006. 'Reframing Development: Global Peasant Movements and the New Agrarian Question.' *Canadian Journal of Development Studies/Revue Canadienne d'études du Développement* 27 (4): 471–483.

———. 2009. 'Food Sovereignty, Social Reproduction and the Agrarian Question.' In *Peasants and Globalisation: Political Economy, Rural Transformation and the Agrarian Question*, edited by A. H. Akram-Lodhi and C. Kay, 288–312. London: Routledge.

———. 2013. *Food Regimes and Agrarian Questions*. Halifax: Fernwood Publishing.

Mehrotra, K. N. 1989. 'Pesticide Resistance in Insect Pests: Indian Scenario.' *Pesticide Research Journal* 1 (2): 95–103.

Mencher, J. P. 1974. 'Conflicts and Contradictions in the "Green Revolution": The Case of Tamil Nadu.' *Economic and Political Weekly* 9 (6–8): 309–323.

Mishra, S. 2009. 'Agrarian Distress and Farmers' Suicides in Maharashtra.' In *Agrarian Crisis in India*, edited by D. N. Reddy and S. Misha. Delhi: Oxford University Press.

Mohanty, M. 1998. 'Social Movements in Creative Society: Of Autonomy and Interconnection.' In *People's Rights: Social Movements and the State in the Third World*, edited by M. Manoranjan, P. N. Mukherji, and O. Törnquist, 63–81. New Delhi: Sage Publications.

Mosse, D. 2001. '"People's Knowledge", Participation and Patronage: Operations and Representations in Rural Development.' In *Participation: The New Tyranny?*, edited by B. Cooke and U. Kothari, 16–35. London: Zed Books.

Mukherji, P. N. 1998. 'The Farmers' Movement in Punjab: Politics of Pressure Groups and Pressure of Party Politics.' *Economic and Political Weekly* 33 (18): 1043–1048.

Muttiah, C. 1970. 'The Agrarian Problem in Thanjavur and the New Agrarian Strategy.' *Indian Journal of Agricultural Economics* 25 (3).

Nanda, M. 2003. *Prophets Facing Backward: Postmodern Critiques of Science and Hindu Nationalism in India*. Rutgers University Press: New Brunswick.

Natrajan, K. 2008. *Panchagavya: A Manual*. Goa: Other India Press.

Navarrete, M. 2009. 'How Do Farming Systems Cope with Marketing Channel Requirements in Organic Horticulture? The Case of Market-gardening in Southeastern France.' *Journal of Sustainable Agriculture* 33 (5): 552–565.

Negi, B. and D. S. Negi. n.d. *Seeds for Life: The Story of Beej Bachao Andolan*. Dehradun: Sri Sai Offset Press.

Nilsen, A. G. 2015. 'For a Historical Sociology of State-society Relations in the Study of Subaltern Politics.' In *New Subaltern Politics: Reconceptualizing Hegemony and Resistance in Contemporary India*, edited by A. G. Nilsen and S. Roy, 31–53. Oxford: Oxford University Press.

Nilsen, A. G. and S. Roy. 2015. 'Reconceptualizing Subaltern Politics in Contemporary India.' In *New Subaltern Politics: Reconceptualizing Hegemony and Resistance in Contemporary India*, edited by A. G. Nilsen and S. Roy, 1–27. Oxford: Oxford University Press.

Oberschall, A. 1973. *Social Conflict and Social Movements*. New Jersey: Prentice Hall, Inc.

———. 1978. 'Theories of Social Conflict.' *Annual Review of Sociology* 4 (1): 291–315.

Omvedt, G. 1993. *Reinventing Revolution: New Social Movements and the Socialist Tradition in India*. London: M. E. Sharpe, Inc.

———. 2005. 'Farmers' Movements and the Debate on Poverty and Economic Reforms in India.' In *Social Movements in India: Poverty, Power and Politics*, edited by R. Ray and M. F. Katzenstein, 179–202. New York: Rowman and Littlefield Publishers.

Organic Uttarakhand. 2012. 'Organic Uttarakhand: Delivering Prosperity through Sustainable Agriculture.' Accessed 3 August 2012. Available at http://www.organicuttarakhand.org/.

Palekar, S. n.d. *The Philosophy of Spiritual Farming*. Amravati, Maharashtra: Amol Subhash Palekar and Amit Subhash Palekar.

Pande, P. N. 1996. *Drudgery of the Hill Women*. New Delhi: Indus Publishing Company.

Pandey, G. 1982. 'Peasant Revolt and Indian Nationalism: The Peasant Movement in Awadh, 1919–1922.' In *Subaltern Studies 1*, edited by R. Guha, 143–197. Oxford: Oxford University Press.

Pastakia, A. 1998. 'Grassroots Entrepreneurs: Change Agents for a Sustainable Society.' *Journal of Organizational Change Management* 11 (2): 157–173.

Patel, R. 2006. 'International Agrarian Restructuring and the Practical Ethics of Peasant Movement Solidarity.' *Journal of Asian and African Studies* 41 (1–2): 71–93.

————. 2013. 'The Long Green Revolution.' *Journal of Peasant Studies* 40 (1): 1–63.

Patnaik, P. and C. P. Chandrasekhar. 1995. 'Indian Economy under "Structural Adjustment".' *Economic and Political Weekly* 30 (47): 1001–1013.

Patnaik, U. 1996. 'Export-oriented Agriculture and Food Security in Developing Countries and India.' *Economic and Political Weekly* 31 (35–37): 2429–2449.

————. 2003. 'Global Capitalism, Deflation and Agrarian Crisis in Developing Countries.' *Journal of Agrarian Change* 3 (1–2): 33–66.

Petras, J. 1999. 'NGOs: In the Service of Imperialism.' *Journal of Contemporary Asia* 29 (4): 429–440.

Pimentel, D. 1993. 'Economics and Energetics of Organic and Conventional Farming.' *Journal of Agricultural and Environmental Ethics* 6 (1): 53–60.

Platteau, J. P. 2004. 'Monitoring Elite Capture in Community-driven Development.' *Development and Change* 35 (2): 223–246.

Preobrazhensky, E. 1965. *The New Economics*. Translated by B. Pearce. Oxford: Clarendon Press.

Pretty, J. 1995. 'Participatory Learning for Sustainable Agriculture.' *World Development* 23 (8): 1247–1263.

————. 2002. 'Social and Human Capital for Sustainable Agriculture.' In *Agroecological Innovations: Increasing Food Production with Participatory Development*, edited by N. Uphoff, 47–57. London: Earthscan.

Pretty, J., J. Morison, and R. Hine. 2003. 'Reducing Food Poverty by Increasing Agricultural Sustainability in Developing Countries.' *Agriculture, Ecosystems and Environment* 95 (1): 217–234.

Pretty, J., R. Ruben, and L. A. Thrupp. 2002. 'Institutional Changes and Policy Reforms.' In *Agroecological Innovations: Increasing Food Production with Participatory Development*, edited by N. Uphoff, 251–260. London: Earthscan.

Putzel, J. 1992. *A Captive Land: The Politics of Agrarian Reform in the Philippines*. New York: Monthly Review Press.

Randhawa, M. S. 1974. *Green Revolution: A Case Study of Punjab*. Delhi: Vikas Publishing House.

Rao, V. M. 2009. 'Farmers' Distress in a Modernizing Agriculture – The Tragedy of the Upwardly Mobile: An Overview.' In *Agrarian Crisis in India*, edited by D. N. Reddy and S. Mishra. Delhi: Oxford University Press.

Rajagopal, A. 2001. 'The Violence of Commodity Aesthetics: Hawkers, Demolition Raids and a New Regime of Consumption.' *Social Text* 19 (3): 91–114.

Ramanjaneyulu, G. V., M. S. Chari, T. A. V. S. Raghunath, Z. Hussain, and K. Kuruganti. 2009. 'Non Pesticidal Management: Learning from Experiences.' In *Integrated Pest*

Management: Innovation-Development Process 1, edited by R. Peshin and A. K. Dhawan, 573–574. Netherlands: Springer.

Randeria, S. 2003. 'Glocalisation of Law: Environmental Justice, World Bank, NGOs and the Cunning State in India.' *Current Sociology* 51 (3–4): 305–328.

Rangan, H. 2000. *Of Myths and Movements: Rewriting Chipko into Himalayan History*. London: Verso.

Ray, R. and M. F. Katzenstein. 2005. 'Introduction: In the Beginning, There was the Nehruvian State.' In *Social Movements in India: Poverty, Power and Politics*, edited by R. Ray and M. F. Katzenstein, 1–31. New York: Rowman and Littlefield Publishers, Inc.

Reddy, D. N. and S. Mishra. 2009a. *Agrarian Crisis in India*. New Delhi: Oxford University Press.

————. 2009b. 'Agriculture in the Reforms Regime.' In *Agrarian Crisis in India*, edited by D. N. Reddy and S. Mishra, 3–43. New Delhi: Oxford University Press.

Reddy, D. N. and C. Upendranadh. 2010. *National Rural Employment Guarantee: Issues, Concerns and Prospects*. New Delhi: Oxfam India.

Riddell, R. 2007. *Does Foreign Aid Really Work?* New York: Oxford University Press.

Roberts, N. S. 2009. 'Impacts of the National Green Corps Program (Eco-Clubs) on Students in India and Their Participation in Environmental Education Activities.' *Environmental Education Research* 15 (4): 443–464.

Rodell, M., I. Velicogna, and J. S. Famiglietti. 2009. 'Satellite-based Estimates of Groundwater Depletion in India.' *Nature* 460: 999–1002.

Rosset, P. and M. A. Altieri. 1997. 'Agroecology verus Input Substitution: A Fundamental Contradiction in Sustainable Agriculture.' *Society and Natural Resources* 10 (3): 283–295.

Rosset, P. and M. Bourque. 2005. 'Lessons of Cuban Resistance.' In *The Earthscan Reader in Sustainable Agriculture*, edited by J. Pretty, 262–367. London: Earthscan.

Roy, B. C. and G. N. Chattopadhyay. 2009. 'Patterns of Synthetic Fertiliser Subsidies and Their Impact on Overuse and Imbalance Use.' In *Subsidising Food Crisis*, edited by B. C. Roy, G. N. Chattopadhyay, and R. Tirado, 3–10. Bangalore: Greenpeace India.

Sahoo, S. 2013. 'Doing Development or Creating Dependency? NGOs and Civil Society in India.' *South Asia* 36 (2): 258–272.

Sainath, P. 2009. 'Neoliberal Terrorism in India: The Largest Wave of Suicides in History.' *CounterPunch*. Retrieved from http://www.counterpunch.org/2009/02/12/the-largest-wave-of-suicides-in-history/.

Samaddar, R. 2008. 'Jayaprakash Narayan and the Problem of Representative Democracy.' *Economic and Political Weekly* 43 (31): 49–58.

Sanyal, K. 2007. *Rethinking Capitalist Development: Primitive Accumulation, Governmentality and Post-Colonial Capitalism*. London: Routledge.

Sarkar, A. and A. Das. 2014. 'Groundwater Irrigation-Electricity-Crop Diversification Nexus in Punjab: Trends, Turning Points, and Policy Initiatives.' *Economic and Political Weekly* 49 (52): 64–73.

Saull, R. 2012. 'Rethinking Hegemony: Uneven Development, Historical Blocs, and the World Economic Crisis.' *International Studies Quarterly* 56 (2): 323–338.

Save, B. 2008. *The Great Agricultural Challenge*. Kolkata: Earthcare Books.

Scheper-Hughes, N. 1995. 'The Primacy of the Ethical: Propositions for a Militant Anthropology.' *Current Anthropology* 36 (3): 409–440.

Scialabba, N. 2000. *Factors influencing organic agriculture policies with a focus on developing countries*. Paper presented at the IFOAM Scientific Conference, 28–31 August 2000, Basel, Switzerland.

Scoones, I. 2015. *Sustainable Livelihoods and Rural Development*. Black Point, Nova Scotia: Fernwood Publishing.

Scott, J. 1985. *Weapons of the Weak: Everyday Forms of Peasant Resistance*. New Haven: Yale University Press.

Sell, S. K. and A. Prakash. 2004. 'Using Ideas Strategically: The Contest between Business and NGO Networks in Intellectual Property Rights.' *International Studies Quarterly* 48 (1): 143–175.

Sen, B. 1974. *The Green Revolution in India*. New York: Wiley.

Sethi, H. 1998. 'Micro-struggles, NGOs and the State.' In *People's Rights: Social Movements and the State in the Third World*, edited by M. Manoranjan, P. N. Mukherji, and O. Törnquist, 408–420. New Delhi: Sage Publications.

Shah, A. 2010. *In the Shadows of the State: Indigenous Politics, Environmentalism, and Insurgency in Jharkhand, India*. Durham: Duke University Press.

Shaner, W. W., P. F. Philip, and W. R. Schmehl. 1981. *Farming Systems Research and Development: Guidelines for Developing Countries*. Boulder, Colorado: Westview Press.

Sharma, M. 2006. 'The Making of Moral Authority: Anna Hazare and Watershed Management Programme in Ralegan Siddhi.' *Economic and Political Weekly* 41 (20): 1981–1988.

Shetty, S. L. 2009. 'Agricultural Credit and Indebtedness: Ground Realities and Policy Perspectives.' In *Agrarian Crisis in India*, edited by D. N. Reddy and S. Mishra, 61–86. New Delhi: Oxford University Press.

Shiva, V. 1989. *Staying Alive: Women, Ecology and Development*. London: Zed Books.

————. 1991. *The Violence of the Green Revolution: Third World Agriculture, Ecology and Politics*. London: Zed Books.

————. 2016. *Earth Democracy: Justice, Sustainability and Peace* (2nd Edition). London: Zed Books.

Sidhu, H. S. 2002. 'A Non-sustainable Model of Agrarian Transformation: The Case Study of Indian Punjab.' In *Sustainable Agriculture, Poverty and Food Security 1*, edited by S. S. Acharya, S. Singh, and V. Sagar, 127–136. New Delhi: Rawat Publications.

————. 2005. 'Production Conditions in Contemporary Punjab Agriculture.' *Journal of Punjab Studies* 12 (2): 197–217.

Sidhu, R. S. and S. Singh. 2004. 'Punjab: Agricultural Wages and Employment.' *Economic and Political Weekly* 39 (37): 4132–4135.

Singh, K. 2009. 'Agrarian Crisis in Punjab: High Indebtedness, Low Returns and Farmers' Suicides.' In *Agrarian Crisis in India*, edited by D. N. Reddy and S. Mishra, 261–284. New Delhi: Oxford University Press.

Singh, K. and S. Kalra. 2002. 'Rice Production in Punjab: Systems, Varietal Diversity, Growth and Sustainability.' *Economic and Political Weekly* 37 (30): 3139–3148.

Singh, M. 2012a. 'Preference for Migrant Agricultural Labour in Punjab.' *Economic and Political Weekly* 47 (29): 27–28.

————. 2012b. 'A Re-election in Punjab and the Continuing Crisis.' *Economic and Political Weekly* 47 (13): 21–23.

Singh, N. and D. S. Kohli. 2005. 'The Green Revolution in Punjab, India: The Economics of Technological Change.' *Journal of Punjab Studies* 12 (2): 285–306.

Singh, N. D. 2010. 'Rural Healthcare and Indebtedness in Punjab.' *Economic and Political Weekly* 45 (11): 22–25.

Singh, R. B. 2000. 'Environmental Consequences of Agricultural Development: A Case Study from the Green Revolution State of Haryana, India.' *Agriculture Ecosystems and Environment* 82 (1–3): 97–103.

Sinha, S., S. Gururani, and B. Greenberg. 1997. 'The "New Traditionalist" Discourse of Indian Environmentalism.' *Journal of Peasant Studies* 24 (3): 65–99.

Singh, S. M. Kaur and H. S. Kingra. 2008. 'Indebtedness among Farmers in Punjab.' *Economic and Political Weekly* 43 (26–7): 130–136.

Singh, S., and S. Bhogal. 2014. 'Punjab's Small Peasantry: Thriving or Deteriorating?' *Economic and Political Weekly* 49 (26–27): 95–100.

Sinha, S. 2003. 'Development Counter-narratives: Taking Social Movements Seriously.' In *Regional Modernities: The Cultural Politics of Development in India*, edited by K. Sivaramakrishnan and A. Agrawal, 286–312. Stanford, California: Standford University Press.

————. 2008. 'Lineages of the Developmentalist State: Transnationality and Village India, 1900–1965.' *Comparative Studies in Society and History* 50 (1): 57–90.

————. 2012. 'Institutional and Policy Aspects of Punjab Agriculture: A Smallholder Perspective.' *Economic and Political Weekly* 47 (4): 51–57.

————. 2014. 'AAP in Punjab: Exploring the Verdict.' *Economic and Political Weekly* 49 (29): 26–27.

————. 2015. 'On the Edge of Civil Society in Contemporary India.' In *New Subaltern Politics: Reconceptualizing Hegemony and Resistance in Contemporary India*, edited by A. G. Nilsen and S. Roy, 225–253. Oxford: Oxford University Press.

Solís, N. P. A. 1995. 'Land and Indigenous Cosmovision.' In *First World, ha ha ha!*, edited by E. Katzenberger, 128. San Francisco: City Lights Books.

Steur, L. 2015. 'Theorizing Thervoy: Subaltern Studies and Dalit Praxis in India's Land Wars.' In *New Subaltern Politics: Reconceptualizing Hegemony and Resistance in Contemporary India*, edited by A. G. Nilsen and S. Roy, 171–201. Oxford: Oxford University Press.

Tanabe, A. 2007. 'Toward Vernacular Democracy: Moral Society and Post-colonial Transformation in Rural Orissa, India.' *American Ethnologist* 34 (3): 558–574.

Tarrow, S. 1998. *Power in Movement: Social Movements and Contentious Politics* (2nd Edition). Cambridge: Cambridge University Press.

Taguchi, Y. 2012. 'Aesthetics of Civil Society: "Fight the Filth" Campaign in Mumbai.' *Economic and Political Weekly* 47 (20): 13–16.

Thakur, J. S., B. T. Rao, A. Rajwanshi, H. K. Parwana, and R. Kumar. 2008. 'Epidemiological Study of High Cancer Among Rural Agricultural Community of Punjab in Northern India.' *International Journal of Environmental Research and Public Health* 5 (5): 399–407.

Thakur, J. S., S. Prinja, D. Singh, A. Rajwanshi, R. Prasad, H. K. Parwana, and R. Kumar. 2010. 'Adverse Reproductive and Child Health Outcomes among People Living near Highly Toxic Waste Water Drains in Punjab, India.' *Journal of Epidemiology and Community Health* 64 (2): 148–152.

Tiwana, N. S., N. Jerath, G. Singh, and R. Singh. 2009. 'Pesticide Pollution in Punjab: A Review.' *Asian Journal of Water, Environment and Pollution* 6 (1): 89–96.

Trauger, A. 2015. 'Seed Sovereignty as Civil Disobedience in Northern India.' In *Food Sovereignty in International Context*, edited by A. Trauger, 106–124. London: Earthscan.

Udayakumar, S. P. 2012. 'The Koodankulam Struggle and the "Foreign Hand".' *Economic and Political Weekly* 47 (12): 11–14.

Uphoff, N. 2002a. *Agroecological Innovations: Increasing Food Production with Participatory Development*. London: Earthscan.

————. 2002b. 'The Agricultural Development Challenges We Face.' In *Agroecological Innovations: Increasing Food Production with Participatory Development*, edited by N. Uphoff, 3–19. London: Earthscan.

Useem, B. 1977. 'Peasant Involvement in the Cuban Revolution.' *Journal of Peasant Studies* 5 (1): 99–111.

Vakil, A. C. 1997. 'Confronting the Classification Problem: Toward a Taxonomy of NGOs.' *World Development* 25 (12): 2057–2070.

Vakulabharanam, V., N. P. Prasad, K. Laxminarayana, and S. Kilaru. 2011. 'Understanding the Andhra Crop Holiday Movement.' *Economic and Political Weekly* 46 (50): 13–16.

Varshney, A. 1995. *Democracy, Development and the Countryside: Urban-Rural Struggles in India*. Cambridge: Cambridge University Press.

Vijayalakshmi, K. and A. V. Balasubramanian. 2004. *Seeds of Plenty, Seeds of Hope*. Chennai: Centre for Indian Knowledge Systems.

Vincent, S. 2004. 'Participation, Resistance and Problems with the "Local" in Peru: Towards a New Political Contract?' In *Participation: From Tyranny to Transformation?*, edited by S. Hickey and G. Mohan, 111–124. London: Zed Books.

Walker, K. 2008. 'Neoliberalism on the Ground in Rural India: Predatory Growth, Agrarian Crisis, Internal Colonisation and the Intensification of Class Struggle.' *Journal of Peasant Studies* 35 (4): 557–620.

Weber, T. 1988. *Hugging the Trees: The Story of the Chipko Movement*. New Delhi: Viking.

Weiner, M. 1962. *The Politics of Scarcity: Public Pressure and Political Response in India*. Chicago: University of Chicago Press.

White, S. 1999. 'NGOs, Civil Society and the State in Bangladesh: The Politics of Representing the Poor.' *Development and Change* 30 (2): 307–326.

Wolf, E. 1969. *Peasant Wars of the Twentieth Century*. New York: Harper and Row.

Yapa, L. 1979. 'Ecopolitical Economy of the Green Revolution.' *The Professional Geographer* 31 (4): 371–376.

Yin, R. K. 2009. *Case Studies Research: Design and Methods* (4th Edition). Thousand Oaks, California: Sage Publications.

Index

Subjects

Names